Mahanadi River

Jay Prakash Bhatt • Ajay Kumar Mishra
Editors

Mahanadi River

The Environmental Challenges and Way Forward

Springer

Editors
Jay Prakash Bhatt
Department of Environmental Science
Sri Sri University
Cuttack, Odisha, India

Ajay Kumar Mishra
South Asia Regional Centre
International Rice Research Institute
Varanasi, Uttar Pradesh, India

ISBN 978-981-96-1047-1 ISBN 978-981-96-1048-8 (eBook)
https://doi.org/10.1007/978-981-96-1048-8

This Springer imprint is published by the registered company Springer Nature Singapore Pte Ltd.
The registered company address is: 152 Beach Road, #21-01/04 Gateway East, Singapore 189721, Singapore

Preface

Freshwater is one of the planet's most precious and limited resources, yet less than 1% of Earth's freshwater is readily available for consumption, primarily found in rivers, lakes, ponds, and other accessible sources. This scarcity makes it highly vulnerable to the pressures of pollution, unsustainable land use, rampant sand mining, encroachment, fishing practices, and river regulation. Wetlands, covering roughly 6% of the Earth's surface, provide vital ecosystem services, supporting nearly 40% of the world's biodiversity. Rivers, as lifelines of these wetland ecosystems, hold immense ecological, economic, cultural, and spiritual significance—particularly in India, where they are revered as sacred entities.

The Mahanadi River, one of the most significant rivers in the Deccan Plateau, stretches over 851 kilometers, draining a vast basin of approximately 141,589 square kilometers across Odisha, Chhattisgarh, Maharashtra, and Jharkhand. Agriculture dominates the land use in the Mahanadi Basin, occupying about 54.27% of the area. Like many of India's major rivers, the Mahanadi serves as a crucial economic, social, and environmental artery for the region. However, it faces escalating environmental threats driven by rapid industrialization, mining, coal-fired power plants, unregulated flow modifications, untreated wastewater, agricultural runoff, and sand mining, among others.

Despite its importance, the Mahanadi has remained underexplored in terms of its socioeconomic value, water quality, biodiversity, and overall environmental health, especially when compared to rivers like the Ganga, Yamuna, Godavari, Krishna, and Cauvery. Unlike the Ganga and Yamuna, which have garnered attention from ecologists, sociologists, civil society organizations, and policymakers, the Mahanadi's challenges have not yet inspired widespread intervention or a focused conservation effort.

To address these gaps, the conference "Mahanadi River: Environmental Challenges and the Way Forward" was held on December 18–19, 2023, at the Sri Sri University Campus. This gathering brought together a diverse array of stakeholders, aiming to spotlight the critical issues facing the Mahanadi and to initiate an awareness campaign for its conservation. The conference emphasized the need for comprehensive baseline data collection, further research, and the development of a clear policy framework for the sustainable management of the river.

It is our honor to present this book, which is based on the papers and discussions from the conference. It offers a multidisciplinary examination of the Mahanadi River's existing and emerging challenges, covering topics such as water quality, biodiversity, fisheries, sand mining, river regulation, land use changes, interstate water conflicts, and the agricultural dynamics within the river basin.

We hope that this book will serve as a valuable resource for researchers, policymakers, and environmental advocates alike, contributing to the ongoing efforts to safeguard the Mahanadi River for future generations.

Cuttack, Odisha, India — Jay Prakash Bhatt
Varanasi, Uttar Pradesh, India — Ajay Kumar Mishra

Acknowledgments

We would like to extend our deepest gratitude to the institutions and organizations whose unwavering commitment to advancing agricultural sustainability has made this book, *Mahanadi River: The Environmental Challenges and Way Forward*, possible. Their continuous encouragement and the resources they provided have been instrumental in shaping our work and this publication.

We sincerely acknowledge the support of Sri Sri University and express our heartfelt thanks to Sri Sri Ravi Shankar, the Indian guru and spiritual leader, whose teachings on sustainable living and spirituality have deeply influenced our approach to regenerative agriculture and holistic environmental stewardship.

A special thanks goes to the Ministry of Water Resources, Government of India, for their steadfast support in promoting sustainable water management practices. Their initiatives and policies have played a crucial role in our understanding of water conservation techniques and their application in regenerative agriculture systems.

We are immensely grateful to Dr. Sudhanshu Singh, Director of the International Rice Research Institute South Asia Regional Centre, for his encouragement and for providing strategic directions. His innovative approach to regenerative agriculture has significantly shaped the work presented in this book.

We also express our sincere appreciation to Prof. Rajita Kulkarni, President, Sri Sri University and Prof. Tej Pratap, Vice-Chancellor of Sri Sri University, for his guidance and leadership in fostering an environment that promotes environmental conservation and sustainable agricultural practices. His vision and support have been a source of inspiration throughout this project.

A special note of gratitude goes to Satyajeet Arya, Assistant Professor, Faculty of Science, and Director at the Sri Sri Centre for Advance Research on Water Resources and Environmental Management, Sri Sri University, Cuttack, Odisha. His invaluable expertise and dedication to water resources and environmental management have significantly contributed to the development of this book.

On a personal note, Prof Jay Prakash Bhatt expresses his heartfelt gratitude to his wife, Mrs. Kalpana Bhatt and his elder brother, Mr. Sita Ram Bhatt for their unwavering support. Dr. Ajay Kumar Mishra, expresses his heartfelt appreciation to his family for their unwavering support and inspiration: his late grandfather, Shri Keshav Mishra, his father, Shri Chandra Bhushan Mishra, his mother, Smt. Kamalawati Mishra, his uncle, Prof. Ram Bhushan Mishra, his father-in-law, Late

Shri Ramanand Mishra, his wife, Sarita Mishra, and his sons, Atharv Mishra and Aarav Mishra. Their love and encouragement have been the cornerstone of his journey.

As you embark on this journey through the pages of *Mahanadi River: The Environmental Challenges and Way Forward*, we hope you find inspiration, knowledge, and a renewed commitment to cultivating a healthier, more resilient world.

Contents

Editors and Contributors

About the Editors

Jay Prakash Bhatt is a distinguished Professor in the field of Environmental Science at Sri Sri University, Cuttack, Odisha. With over two decades of research and teaching experience, Dr. Bhatt has been a driving force in the areas of aquatic ecology, water quality, biodiversity, and environmental management. His vast academic background includes a bachelor's and master's degree from Hemwati Nandan Bahuguna (HNB) Garhwal University, followed by a long association with the Centre for Interdisciplinary Studies of Mountain & Hill Environment and the Department of Environmental Studies at the University of Delhi.

Dr. Bhatt has led numerous environmental impact assessments, biodiversity studies, and carrying capacity studies related to river valley projects, where he served as Principal Investigator and Co-Principal Investigator. His research has been published widely in both national and international journals, and he is known for his contributions to the understanding of fish and fisheries, wildlife, and the broader ecological dynamics of river ecosystems, particularly the Mahanadi River Basin. Dr. Bhatt's work also extends to environmental management and conservation, playing a significant role in shaping policy and practical solutions for ecological sustainability.

As an accomplished academic and researcher, Dr. Bhatt has collaborated on various multidisciplinary projects that span the intersection of ecological, social, and economic dimensions. His commitment to fostering sustainable environmental practices is reflected in his teaching and mentoring of students, as well as his active participation in scientific communities.

Ajay Kumar Mishra is a highly accomplished Senior Associate Scientist in the field of Soil Science, currently working with the International Rice Research Institute (IRRI). With expertise in enhancing rice-based agri-food systems through improved soil nutrient management and Agri-relevant ICT, Mishra is at the forefront of strategic and adaptive research. Mishra's primary focus lies in the development and implementation of the Rice Crop Manager tool, an ICT-based platform that delivers context-specific rice farming management technologies. Collaborating with public and private partners, he actively contributes to the rapid deployment of actionable and tailored farming solutions. His work is particularly targeted toward small-scale farmers in India, where the application of these technologies helps improve profitability and productivity in rice and rice-based agri-food farming. Apart from his involvement in ICT-based solutions, Mishra is dedicated to developing low-cost innovative technologies for climate-smart agriculture and ensuring their adoption by the farming community. He has also played a pivotal role in the capacity development of extension agents and scholars, imparting knowledge and skills to drive positive change in agricultural practices. Mishra's research contributions are significant, as evident from his authorship and co-authorship of over 30 scientific papers in peer-reviewed journals and book chapters. His expertise, dedication, and collaborative approach make him an invaluable asset in advancing sustainable and efficient agricultural practices in the context of rice farming.

Contributors

Satyajeet Arya Environmental Science, Faculty of Science, Sri Sri University, Cuttack, Odisha, India

Ashok Kumar Behera Department of Environmental Sciences, Faculty of Agriculture, Sri Sri University, Odisha, India

Jay Prakash Bhatt Faculty of Science, Sri Sri University, Cuttack, Odisha, India

Rachna Chandra Department of Environmental Sciences, Faculty of Agriculture, Sri Sri University, Cuttack, Odisha, India

Anannya Das Environmental Science, Faculty of Science, Sri Sri University, Cuttack, Odisha, India

Pratik Kumar Dash P.G. Department of Environmental Sciences, Sambalpur University, Sambalpur, Odisha, India

Prava Kiran Dash Department of Soil Science and Agricultural Chemistry, Institute of Agricultural Sciences (IAS), Siksha 'O' Anusandhan Deemed to be University, Bhubaneswar, Odisha, India

Regional Research and Technology Transfer Station, Dhenkanal, Odisha, India

Subrat Kumar Das P.G. Department of Environmental Sciences, Sambalpur University, Sambalpur, Odisha, India

Tejaswini Dash P.G. Department of Environmental Sciences, Sambalpur University, Sambalpur, Odisha, India

Karuna Ekka P.G. Department of Environmental Sciences, Sambalpur University, Odisha, India

Narayan Jena Faculty of Arts, Communication, and Indic Studies, Sri Sri University, Cuttack, Odisha, India

Prasanna Kumar Orissa Remote Sensing Application Centre, Bhubaneswar, Odisha, India

Ajay Kumar Mishra International Rice Research Institute, South Asia Regional Centre, Varanasi, Uttar Pradesh, India

Antaryami Mishra Department of Soil Science and Agricultural Chemistry, Institute of Agricultural Sciences (IAS), Siksha 'O' Anusandhan Deemed to be University, Bhubaneswar, Odisha, India

Sarita Mishra Faculty of Science, Sri Sri University, Cuttack, Odisha, India

Malaya Ranjan Mahananda P.G. Department of Environmental Sciences, Sambalpur University, Sambalpur, Odisha, India

Utpal Mohanta Environmental Science, Faculty of Science, Sri Sri University, Cuttack, Odisha, India

Bidut Prava Mohanty P.G. Department of Zoology, Panchayat College, Bargarh, Odisha, India

Vijaya Lakshmi Mohanty Human Resoure Development Centre (HRDC), Sri Sri University, Cuttack, Odisha, India

Hitanshu Shekhar Mohanta School and Mass Education Department, Government of Odisha, Bhubaneswar, Odisha, India

Sanjaya Kumar Pattanayak P.G. Department of Environmental Sciences, Sambalpur University, Sambalpur, Odisha, India

Abheepsha Priyadarshini Pradhan Environmental Science, Faculty of Science, Sri Sri University, Cuttack, Odisha, India

Narad Rajesh P.G. Department of Environmental Sciences, Sambalpur University, Sambalpur, Odisha, India

Rishab Ray Faculty of Architecture, Sri Sri University, Cuttack, Odisha, India

Rakesh Roshan Sahu P.G. Department of Environmental Sciences, Sambalpur University, Sambalpur, Odisha, India

Sheetal Sharma International Rice Research Institute (IRRI), South Asia Regional Centre, Varanasi, Uttar Pradesh, India

Krishna Pratap Singh Institute of Indian and Foreign Languages, MGM University, Chhatrapati Sambhajinagar, Maharashtra, India

Snigdha Snehanjali Faculty of Science, Sri Sri University, Cuttack, Odisha, India

Sonal Tiwari School of Planning and Architecture, Bhopal, Madhya Pradesh, India

Manas Ranjan Sahoo International Rice Research Institute (IRRI), South Asia Regional Centre, Varanasi, Uttar Pradesh, India

Mahanadi River: Environmental Challenges and Needs for its Sustainability

1

Sanjaya Kumar Pattanayak, Narad Rajesh, Pratik Kumar Dash, Subrat Kumar Das, and Karuna Ekka

Abstract

The Mahanadi River Basin is a peninsular basin that drains into the Bay of Bengal. It covers an area of 139681.51 sq. km and flows through the states of Chhattisgarh, Odisha, Maharashtra, Jharkhand, and Madhya Pradesh. In the past 20 years, different environmental sectors of the basin have degraded because of rapid development in the area in terms of hydropower projects, road and railway projects, industrial projects, mining projects, and the expansion of cities. The geology, physiography, land use/land cover changes, water quality, and developmental activities in the Mahanadi basin, have been critically examined in this discussion. The different impacts likely to occur in the basin have been unveiled and measures required to ameliorate them to safeguard the ecosystem of the basin have been deciphered.

Keywords

Mahanadi River Basin · Environmental challenges · Sustainability

1.1 Introduction

A river basin, also known as a watershed or drainage basin, is a geographical area drained by a river and its tributaries. It functions as a fundamental hydrological unit, collecting precipitation and channeling it through a network of rivers, streams, and other watercourses toward a common outlet, such as an ocean, sea, lake, or another

S. K. Pattanayak (✉) · N. Rajesh · P. K. Dash · S. K. Das · K. Ekka
P.G. Department of Environmental Sciences, Sambalpur University, Sambalpur, Odisha, India
e-mail: skpattanayak@suniv.ac.in; pratikkumardash7@suniv.ac.in

J. P. Bhatt, A. K. Mishra (eds.), *Mahanadi River*,
https://doi.org/10.1007/978-981-96-1048-8_1

river. River basins are vital for sustaining ecosystems, supporting human livelihoods, and driving economic development. They encompass a wide range of land uses, from forests and agricultural lands to urban areas, and are characterized by complex interactions between natural processes and human activities (Chandrakar and Dhuria 2020). India is a diverse country. The diversity is even seen in its river system. The river systems of India can be categorized into three types: the Himalayan river system, which includes perennial rivers such as Ganga, Indus, and Brahmaputra; the peninsular river systems that are entirely Monsoon-fed, such as Godavari, Krishna, Cauvery, and Mahanadi; and lastly, the west-facing peninsular river system, which includes rivers such as Narmada and Tapi.

The Mahanadi (maha, mighty; nadi, river), which originates at 442 m above mean sea level (MSL) near Pharsiya village in the Dhamtari district of Chhattisgarh, has a catchment area of 139681.51 sq. km, accounting for nearly 4.28% of the total geographical area of the country, and is the eighth largest basin in India. The geographical extent of the basin varies from 80°28′ to 86°43′ east longitudes and 19°8′ to 23°32′ north latitudes. It is physically bounded by Central India hills in the north, Eastern Ghats in the south and east, and the Maikal hill range in the west. The general shape of this basin is circular, with a maximum length and width of 587 km and 400 km, respectively. The basin covers five major states: Chhattisgarh, 73214.52 sq. km (52.42%); Odisha, 6587.28 sq. km (47.14%); Maharashtra, 322.38 sq. km (0.23%); Madhya Pradesh, 151.68 sq. km (0.11%), and Jharkhand, 145.66 sq. km (0.1) (CWC 2014).

1.2 Structure of the River Basin

The structure of the Mahanadi River Basin can be understood in terms of its physiography, geology, and major tributaries.

1.2.1 Physiography

The terrain of the basin varies, with elevations ranging from 5 m in the coastal lowlands to 1321 m in the northern hills. The Mahanadi Basin is divided into three subbasins – upper Mahanadi, middle Mahanadi, and lower Mahanadi. The lower basin is the largest subbasin, followed by the middle and upper subbasins. The upper basin is characterized by the rugged topography of the Eastern Ghats and the Chhattisgarh plains; it is drained by the main river and several tributaries, including the Seonath, Hasdeo, and Jonk rivers. The drainage pattern is predominantly dendritic and influenced by the underlying geology and topography. The middle basin extends into parts of eastern Chhattisgarh and western Odisha. As the river moves eastward into the Odisha plains, the topography becomes gentler, with wide floodplains. This section is heavily influenced by the deposition of sediments resulting in extensive alluvial plains. The Mahanadi River cuts through Eastern Ghats in this region, creating a gorge near Hirakud dam, one of the largest dams in the world. The drainage pattern here is complex, with the river receiving significant contributions from tributaries such as Tel and Mand. Meanwhile, the lower basin includes the eastern parts of Odisha, stretches toward the Bay of Bengal, which includes the Mahanadi Delta, and is characterized by low-lying areas prone to flooding. The river splits into multiple distributaries as it approaches the Bay of Bengal, including the Kathajodi, Kuakhai, Devi, and Paika rivers. The deltaic region has a complex network of waterways, swamps, and estuaries influenced by tidal activities (CWC 2014).

1.2.2 Geology of the Mahanadi River Basin

Panigrahy and Raymahashay (2005) reported that the upper catchment of the basin is dominated by sedimentary rocks of the Chhattisgarh Basin, which consists mainly of limestone with subordinate amounts of sandstone and shale (Fig. 1.1). The limestone contains predominantly calcite, dolomite, and ankerite with minor quantities of chlorite, quartz, feldspar, and kaolinite. The shale consists of quartz, feldspar, dolomite, calcite, illite, kaolinite, and hematite. The sandstone consists of quartz, feldspar, and glauconite. Wani (2008) reported the average chemical composition of granite and gneiss from the Bastar Craton, which chiefly consists of SiO_2 (71%), followed by approximately 14% Al_2O_3, 3–4% K_2O and Na_2O, and 2% Fe_2O_3 and

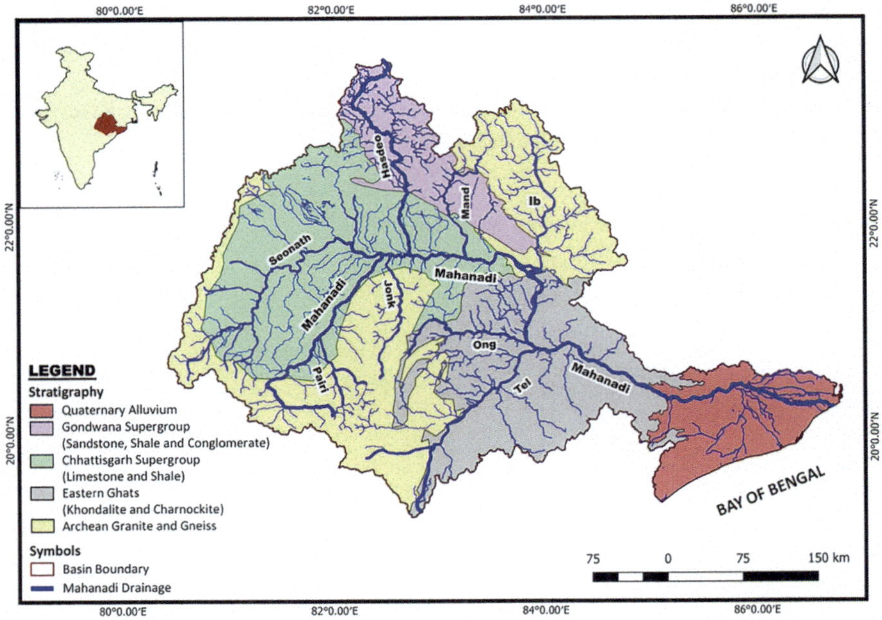

Fig. 1.1 Geological map of the Mahanadi River Basin. (Modified after Bastia et al. 2020)

CaO. The basin is rich in deposits of coal, gold, pyrite, bauxite, quartz, dolomite, lime, zinc, and mica (Chakrapani and Subramanian 1993a, b). These mineralized zones host heavy metals that are found in the river bed of Mahanadi according to Chakrapani and Subramanian (1990). Carbonate rock weathering controls the water chemistry. The bicarbonate constitutes nearly 50% of the dissolved flux to the Bay of Bengal. The water composition is in equilibrium with that of dolomite, kaolinite, and chlorite (Chakrapani and Subramanian 1993a, b). Mineralogically, the dominant constituent of the sediment in the Mahanadi River Basin is quartz, trailed by orthoclase and plagioclase. Most of the sediments contain intermediate amounts of calcite and dolomite. Among the clay minerals, the dominant constituents are kaolinite and illite, along with smectite. The source of the sediment is mostly felsic igneous rock, with some contributions from mafic and carbonate rocks (Bastia et al. 2020).

1.2.3 Major Tributaries

The Mahanadi River has many tributaries (Table 1.1) and distributaries. The main stream, Mahanadi, which has a wide channel, originates at Maikal, whereas the longest stream, Seonath, originates from Gadchiroli in Maharashtra.

In the lower reach of the basin, there are many distributaries of the Mahanadi River, viz.:

(i) Kathajodi: The largest distributary of Mahanadi, which separates from the river before the city of Cuttack and drains into the Bay of Bengal in Puri district.
(ii) Birupa: Birupa separates from Mahanadi and joins the Brahmani River near Krushnanagar and drains into the Bay of Bengal near Dhamra.

Table 1.1 List of the major tributaries of Mahanadi

Tributary	Length of the river (km)	Catchment area (km^2)	Origin	States
Left bank tributaries				
Seonath river	365.76	30,761	Gadchiroli district, Maharashtra	Chhattisgarh
Hasdeo river	340.25	9856	Maikal hills	Chhattisgarh
Mand river	260.01	5200	Surguja district, Chhattisgarh	Chhattisgarh
Ib river	269.25	12,447	Hills near Pandrapet	Odisha
Right bank tributaries				
Ong river	260.01	5128	Northern outskirts of Khariar hills, Odisha	Odisha
Jonk river	197.92	3484	Khariar hills	Chhattisgarh
Tel river	307.23	22,818	Nabarangpur	Odisha

(iii) Kuakhai, Devi, and Surua: Streams that fall into the Bay of Bengal after entering the Puri district.
(iv) Kushabhadra, Bhargavi, and Daya: These three distributaries branch off downstream from the delta apex; Bhargavi and Daya drain into Chilika Lake, while Kushabhadra drains into the Bay of Bengal.
(v) Paika, Chitroptala, Genguti, and Lun: These distributaries are small distributaries of Mahanadi. The Mahanadi proper enters the sea via several channels near Paradeep at False Point, Jagatsinghpur.

1.3 Water Chemistry

The water chemistry of the Mahanadi River Basin is shaped by a combination of geological formations, climatic conditions, land use patterns, and human activities. The basin receives water from various tributaries, each contributing distinct chemical signatures based on the surrounding geology and land practices. The confluence of diverse geological formations, such as Precambrian rocks, Gondwana sediments, and alluvial deposits, plays a significant role in determining the baseline water chemistry. Studies on the physicochemical parameters of water and heavy metals in the Mahanadi River Basin conducted by various researchers in the last decade are presented in Tables 1.2 and 1.3, respectively.

A study conducted by Das (2022) in different cities in the Mahanadi River Basin categorized the cities into three categories. First is Cluster-1 (low pollution), which consists of Hirakud, Brajrajnagar, Sundergarh, Sonepur, Tikarpada, Narsinghpur, Jharsuguda, Ulunda, Boudh, and Tigiria. Second is Cluster-2 (moderately polluted), which consists of Sambalpur, Sonepur, Brajrajnagar, Dhama, Athamallik, and Choudwar. Lastly, third is Cluster-3 (heavy pollution), which is mainly Cuttack and Paradeep; it consists of the downstream part and is the most polluted due to high urbanization, intensive agriculture, and the presence of industries. Home sewage from Paradeep Township, two fertilizer factories, and other small-scale industries discharges waste into the Mahanadi estuary. In addition, fishing harbor operations in estuaries also cause deterioration in water quality. Apart from a good amount of solid waste from phosphatic fertilizer plants situated in the estuary, Atharbanki creek also receives most of the municipal sewage from Paradeep Township. These wastes are allowed to join the sewers' untreated bodies and create contamination. The results also revealed that the discharge of sewage into the Mahanadi estuary should not be performed considering the present state of overflow (Dixit et al. 2013). The Mahanadi River at the Cuttack is highly polluted compared with the other districts in the basin. The water in this region is unsuitable for human consumption (Das and Panda 2010). The Mahanadi River water in the Hirakud region is also classified as having poor water quality at all sampling locations, making it unfit for human consumption (class D), as per the National Sanitation Foundation (NSF) water quality index. However, water can be

Table 1.2 Physicochemical parameters of water from various tributaries of the Mahanadi River Basin

Sl No.	Location/river	State	Temperature (°C) Mean	Temperature (°C) SD	pH Mean	pH SD	Conductivity (µS/cm) Mean	Conductivity (µS/cm) SD	DO (mg/L) Mean	DO (mg/L) SD	BOD (mg/L) Mean	BOD (mg/L) SD	Total coliforms (MPN/100 mL) Mean	Total coliforms (MPN/100 mL) SD	TDS (mg/L) Mean	TDS (mg/L) SD	Nitrates (mg/L) Mean	Nitrates (mg/L) SD	Sulphate (mg/L) Mean	Sulphate (mg/L) SD	Citations
1	Seonath	Chhattisgarh	–	–	7.76	0.39	624.06	373.24	7.01	2.87	23.46	15.58	–	–	258.96	135.65	2.97	2.94	7.30	7.66	Mise and Mujawar (2017)
2	Hasdeo	Chhattisgarh	24.80	–	8.08	0.12	299.70	17.10	9.80	0.23	7.00	1.00	–	–	202.00	7.80	0.24	0.01	44.42	3.20	Bhaskar et al. (2020)
3	IB	Odisha	31.50	0.12	7.69	1.87	–	–	–	–	3.36	0.03	–	–	85.00	404.00			81.06	97.68	Sahoo et al. (2021), Pradhan et al. (2017)
4	Jonk	Chhattisgarh	–	–	7.48	0.21	865.00	78.98	5.28	0.92	1.80	0.35	–	–	657.77	97.65	2.46	0.63	82.88	5.98	Sarwa et al. (2019)
5	Mand	Chhattisgarh	–	–	6.76	0.43	290.32	196.42	–	–	–	–	–	–	145.32	98.31	16.18	28.13	12.34	14.63	Ahmed et al. (2022)
6	Ong	Odisha	–	–	7.20	0.08	720.28	25.90	1.58	0.17	7.13	0.53	–	–	178.75	8.73	9.00	0.32	–	–	Gadria et al. (2022)
7	Mahanadi	Chhattisgarh/ Odisha	–	–	7.80	0.05	–	–	7.50	0.14	1.73	0.34	–	–	6656.45	3007.19	1.95	0.41	190.45	84.68	Das (2023)
8	Mahanadi	Odisha	28.70	1.01	7.48	0.06	0.81	0.11	7.19	0.35	32.30	8.41	479.00	21.00	268.60	29.22	–	–	–	–	Kar et al. (2010)
9	Mahanadi estuary	Odisha	–	–	8.10	0.00	–	–	6.20	0.50	1.50	0.30	–	–	–	–	2.80	1.60	–	–	Mishra et al. (2018)
10	Mahanadi shoreline	Odisha	–	–	8.00	0.30	–	–	6.40	0.30	1.60	0.60	–		–	–	8.60	5.80	–	–	Mishra et al. (2018)
Water quality class parameter																					
Water quality class	**A**		–		**6.5–8.5**		–		**6**		**2**		**50**		**500**		**20**		**400**		**BIS water quality standards**
	B		–		**6.5–8.5**		–		**5**		**3**		**500**		–		–		–		
	C		–		**6.5–8.5**		–		**4**		**3**		**5000**		**1500**		**50**		**400**		
	D		–		**6.5–8.5**		**1000**		**4**		–		–		–		–		–		
	E		–		**6.5–8.5**		**2250**		–		–		–		**2100**		–		**1000**		

Table 1.3 Heavy metal status of the tributaries of the Mahanadi River Basin

Sl No.	Location/river	State	Copper (mg/L)		Iron (mg/L)		Manganese (mg/L)		Zinc (mg/L)		Lead (mg/L)		Cadmium (mg/L)		Boron (mg/L)		Mercury (mg/L)		Chromium (mg/L)		Citations
			Mean	SD	Mean	SD	Mean	SD	Mean	SD	Mean	SD	Mean	SD	Mean	SD	Mean	SD	Mean	SD	
1	IB	Odisha	0.415	0.688	–	–	–	–	0.032	0.210	BDL	–	0.001	0.004	0.023	0.014	BDL		0.031	0.012	Sahoo et al. (2021), Pradhan et al. (2017)
2	Jonk	Chhattisgarh	0.049	0.005	0.225	0.049	0.047	0.006	0.013	0.002	0.039	0.027	BDL		BDL		BDL		BDL		Sarwa et al. (2019)
3	Hasdeo	Chhattisgarh	0.014	0.004	1.717	0.058	0.257	0.026	–	–	0.006	0.004	0.001	0.001	–	–	–	–	–	–	Dhanesh et al. (2014), Vaishnav et al. (2014)
4	Mahanadi	Odisha	0.003	0.002	0.154	0.079	0.016	0.015	0.061	0.089	0.008	0.003	0.001	0.001	–	–	0.008	0.004	0.004	0.002	Sundaray et al. (2012)
5	Mahanadi	Odisha	0.780	0.061	0.150	0.002	–	–	0.070	0.002	0.009	0.000	0.001	0.000	–	–	–	–	–	–	Paital et al. (2023)
Water quality class		*Water quality class parameter*																			
		A	**1.5**		**0.3**		**0.5**		**15**		**0.1**		**0.01**				**0.001**		**0.5**		BIS water quality standards
		B																	**0.5**		
		C	**1.5**		**50**				**15**		**0.1**		**0.01**						**0.5**		
		D																			
		E													**2**						

used for irrigation. The microbiological and physicochemical characteristics of the Mahanadi River water in and around Hirakud are also impacted by activities linked to mining, industry, farming, and irrigation (Kar et al. 2010). The heavy metal concentrations in the Mahanadi River Basin found at various sites are iron (Fe), chromium (Cr), copper (Cu), lead (Pb), and nickel (Ni). These heavy metals are the result of various anthropogenic activities, such as industrialization and mining activities. Arsenic (AS), Cadmium (Cd), mercury (Hg), and zinc (Zn) were found to be well within the acceptable limits prescribed by the World Health Organization (WHO) and Bureau of Indian Standards (BIS) (Hussain et al. 2020). Factor analysis (Das and Tripathy 2019) demonstrated that untreated or partially treated municipal sewage and home and industrial wastewater were the main factors causing variations in water quality. Under high-flow conditions, the major source of contaminants is urban land use. In contrast, during periods of low flow, agricultural pollutant sources govern the quality of water.

1.4 Land Use and Land Cover

The land use/land cover (LULC) maps for the years 2005–2006 and 2017–2018 revealed 23 LULC classes in the basin (Fig. 1.2). Most of the basin is covered with agricultural land, accounting for 49.59% of the total basin area. Among these total agricultural areas, the crop area accounts for 35.98%, and the fallow land accounts for 13.61% of the basin. Except in the districts Durg and Raipur and coastal plains, the basin has a fair area under forests, accounting for 25.79% of the total area. A total of 3.81% of the basin area is covered by waterbodies, which include reservoirs, lakes, and rivers, among others. The two largest waterbodies in the basin are the Hirakud reservoir and Chilika Lake, which can be easily spotted on the LULC map (India-WRIS, 2014).

Anthropogenic activities are occurring daily in the Mahanadi River Basin. As a result, the total forest cover area was reduced by 5.71% of the total area of the basin from 1972 to 2003. This is followed by a reduction in barren land areas of 0.64% and an increase in surface waterbody areas of 0.47%, built-up land areas of 0.22%, riverbed areas of 0.11%, and, most prominently, agriculture areas of 5.55%. This indicates that the total forest cover and barren land have decreased at the expense of the increases in waterbodies, riverbeds, agriculture, and built-up land over a span of 30 years, i.e., from 1972 to 2003 (Dadhwal et al. 2010). Naha et al. (2021), using the LUH2 (Land-Use Harmonization 2) dataset, predicted that by the end of 2021, there would be a remarkable increase in the cropland cover in the Mahanadi River Basin, which was approximately 23.3%, with a very large loss in forest cover of as much as 22.65% compared with that in the baseline year 2005 (Naha et al. 2021). The data presented in Table 1.4 suggest that from 2005 to 2017, there was a decrease in current fallow (42%), littoral swamp (0.85%), shifting cultivation (21.20%), wasteland (9.33%) and waterbodies (2.09). These are indications of large-scale developmental activities in the basin for which large amounts of land area are needed.

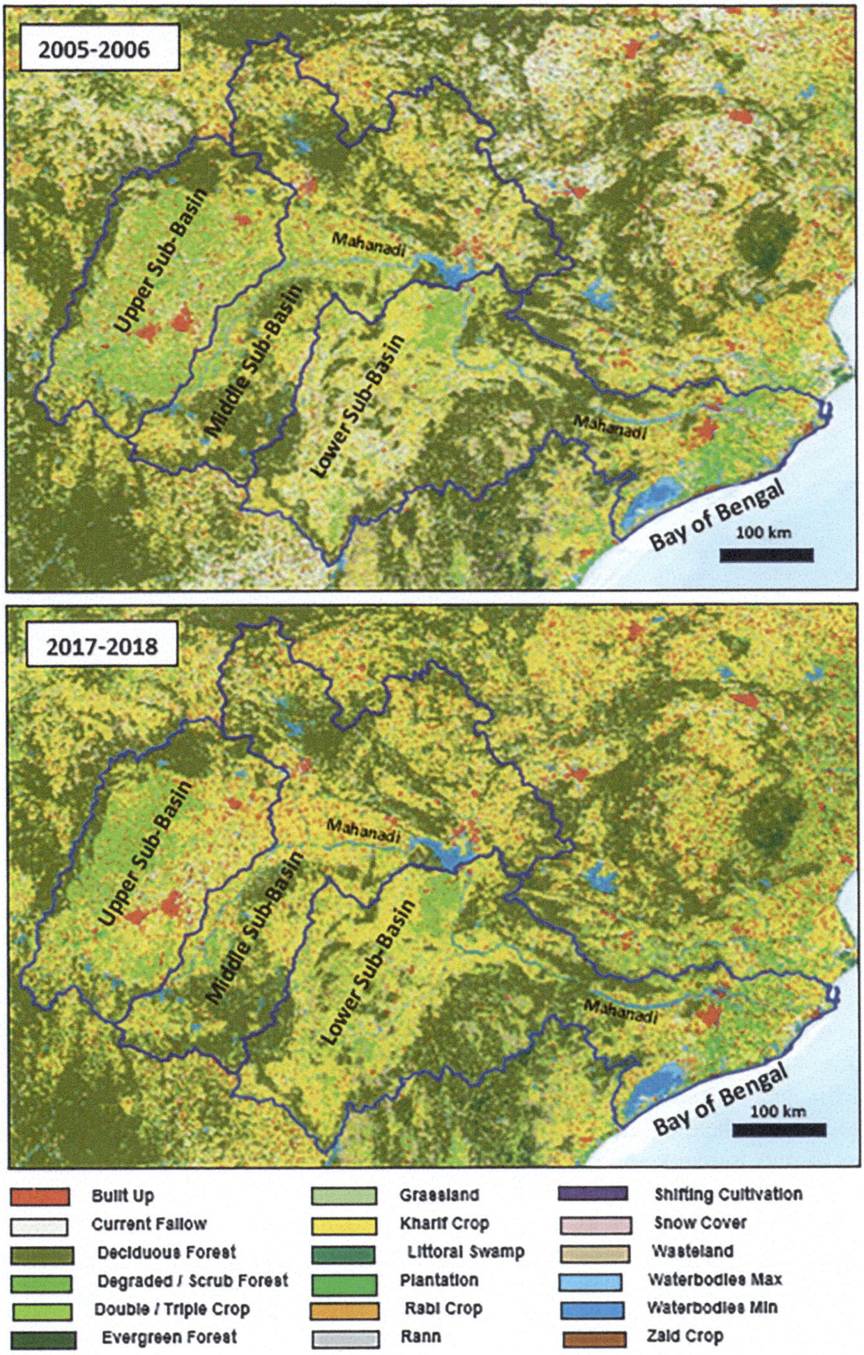

Fig. 1.2 LULC of the Mahanadi River Basin for the years 2005–2006 and 2017–2018 (indiawris.gov.in)

Table 1.4 LULC changes from 2005–2006 to 2017–2018 in the Mahanadi River Basin

	2005–2006		2017–2018		Change (%)	
Class	Area sq. km	Area in %	Area sq. km	Area in %		
Build-up	78529.83	2.47	95673.94	3.01	21.83	Increase
Net shown area (Kharif crop+ Rabi crop+ Zaid crop+ double/triple crop)	1143603.35	35.98	1362514.66	42.88	19.14	Increase
Current fallow	432516.29	13.61	250719.96	7.89	42.03	Decrease
Plantation	92183.77	2.90	93331.55	2.94	1.24	Increase
Forests (evergreen forest+ deciduous forest+ scrub forest)	727420.45	22.89	728111.49	22.92	0.09	Increase
Littoral swamp	4135.82	0.13	4100.32	0.13	0.85	Decrease
Grassland	21934.67	0.69	23004.99	0.72	4.87	Increase
Shifting cultivation	1590.33	0.05	1253.17	0.04	21.20	Decrease
Wasteland	499171.43	15.71	402664.65	12.67	9.33	Decrease
Rann	15717.20	0.49	15872.77	0.50	0.98	Increase
Waterbodies	121151.47	3.81	118608.85	3.73	2.09	Decrease
Snow cover	40460.47	1.27	81451.53	2.56	101.31	Increase

Source: National Remote Sensing Centre, indiawris.gov.in

1.5 Anthropogenic Impacts on the Basin

1.5.1 Man-Made Water Resources in the Mahanadi River Basin

The Central Water Commission (CWC) lists a total of 253 dams and 24 barrages/weirs/anicuts in the Mahanadi River Basin. Among these projects, 74 are large- or medium-sized irrigation projects with potential gross irrigable areas of 15.4 lakh Ha in Odisha and 17.4 lakh Ha in Chhattisgarh, which is a total of 32.8 lakh Ha. This area comprises 40% of the river basin's 82.3 Lakh Ha gross cropping area. In addition to reservoirs, the river basin comprises 3839 tanks that are recognized for their role in surface water storage. In the states of Chhattisgarh and Odisha, these tanks represent a vital system that fulfils various nondrinking household purposes alongside irrigation requirements. The recent developments regarding surface water resources are therefore hot topics of disagreement between these two riparian states. This dispute is related to 13 barrages and diversion weirs along Mahanadi, including the Arpa Bhaisajhar Barrage Project, and the 7 pick-up weirs that are under construction in Chhattisgarh state. Among these, two projects under construction, namely, the Pairy-Mahanadi Intra State Link Project and the Tandula Reservoir Augmentation Scheme on the Mahanadi River Project, are also under Odisha scrutiny. Both are meant for industrial water supply. The study carried out by Joy et al. (2021) on the issue of the water conflicts on eight new barrages and one anicut shows that the annual requirement proposed to be allocated for industries is approximately 1258.16 MCM. These barrages are likely to supply water to more than 45 power plants spread across the 3 districts of Chhattisgarh. Water resources in the Chhattisgarh section of the Mahanadi Basin developed far earlier than those in Odisha, with numerous sizable dams built in the pre-independence era (although much less in capacity than Hirakud). The Maniyari (1930) and Kharang (1931) dams were constructed on the Seonath, the main tributary of the Mahanadi, and the Tandula Tank (1920) in the hills where the Mahanadi originates. Hirakud, Minimata Bango, the Mahanadi Reservoir Complex, and numerous other small- and medium-sized projects are among the other notable projects.

Among the significant large- and medium-scale projects pertaining to Mahanadi and its tributaries are as follows:

1. Hirakud dam: This dam is one of the most prestigious large-scale multipurpose river valley projects in independent India and was completed in 1957. The length of the Masonry dam is 1248 m; the gross storage capacity is 5818 Mm^3, and the height is 61 m. It falls under one of the largest reservoirs in the area built to generate power, facilitate irrigation, and enable flood control. The reservoir, which forms one of the largest artificial lakes in Asia, stretches from the Lamdungri hills to the Chandili Dungri hills. The installed capacity of the hydroelectric plant at the dam is 307.5 MW.
2. Hasdeo Bango: It is situated in the Korba district of Chhattisgarh approximately 70 km from Korba. This reservoir is a multipurpose storage reservoir and the highest dam in the Mahanadi Basin. It is located on the Hasdeo river, a tributary of Mahanadi. It has three 40-MW powerhouse units and one 20-MW firm power unit.
3. Ravishankar Sagar: This dam is located approximately 92 km south of the main city of Raipur and was built in 1978. The multipurpose reservoir provides water for the Bhilai Steel Plant's industrial needs and for irrigation and hydropower purposes. The maximum depth of this water reservoir is approximately 32 m, with the capacity to hold 909 million cubic meters of water. The sediments from this dam are organic matter-poor and nutrient-poor.
4. Dudhawa reservoir: This reservoir is located near the village Dudhawa, approximately 21 km west of Sihawa, which is also near the origin of Mahanadi and 29 km east of Kanker. This project was constructed to deliver water to the project complex of Ravishankar Sagar so that the irrigation potential can be extended. It was brought into operation between 1963 and 1964. This earthen dam has a maximum height of 24.53 m and is 2906.43 m in length.
5. Sondur reservoir: This reservoir was constructed in 1988 across the Sondur river, a tributary of Mahanadi. The catchment is predominantly in the Odisha district of Koraput and in the Chhattisgarh district of Dhamtari. This project aims at enhancing irrigation potential as well as making available water to the Ravishankar Sagar project complex by utilizing the Dudhawa Reservoir.

The large hydroelectric projects in Odisha on the Mahanadi River are the −72 M.W. Chiplima Hydroelectric Project and 235.5 M.W. Hirakud Dam Hydroelectric Project at Burla. The Hirakud dam project inundated a large part of the area where the reservoir achieved the Ramsar status very recently. The Hirakud dam is the longest earth dam in the world and has a large command area irrigating the Bargarh and Sambalpur districts.

1.5.2 Mining and Industrial Activities in the Basin

The Mahanadi Basin has a large reserve of coal. In the energy sector, coal is the primary fuel. The rapid industrialization of the country is forcing state governments to continue coal mining at the cost of environmental degradation. Almost 48 elements are present in coal ash, including radioactive elements and toxic heavy metals. Odisha has more than 24% of the country's coal reserve, whereas Chhattisgarh has 16% coal reserve. In addition to causing considerable ecological damage by removing the forests and topsoil of the basin, coal mining has caused heavy contamination of the river. Coal-fed power plants are also the chief source of emissions of greenhouse gases. They heavily contaminate nearby farms, agricultural fields, and waterways. The Mahanadi Basin abounds in natural resources, has adequate electricity, and therefore has an ideal industrial climate. The major industries present in the basin are Indian Aluminium Co. Ltd.; Smelter and Captive Power Plant at Hirakud; Power Generation Corporation (OPGC) at Banharpali (near Brajrajnagar); Tata Refractories at Brajrajnagar; L & T Cement at Jharsuguda; CPP of ICCL at Choudwar, Paradeep; Oswal Ltd. at Paradeep; Paradeep Phosphates Ltd.; cement factory at Sundargarh; and paper mill near Cuttack. The other industries highly dependent upon agricultural products include the sugar, textile, and oil mill industries. The other industrial sectors are coal mining, iron mining, and manganese mining. Major companies such as Hindalco at Hirakud, Cosboard Industries at Jagatpur, SMV Beverages Private Limited at Cuttack, and the paper industry at Brajarajnagar have been releasing untreated effluents into the Mahanadi River, pushing it toward an impending disaster.

1.6 Impacts of Developmental Activities on the Mahanadi River Basin

1.6.1 Pollution

Various stretches of the Mahanadi River were rated as "polluted river stretches" by the Central Pollution Control Board (CPCB) because of their high biochemical oxygen demand (BOD) values, indicating acute levels of organic pollution. In fact, the values of BOD mostly exceeded the criterion value of 3 mg/L in some stretches and underlined very severe levels of pollution. According to Odisha State Pollution Control Board (OSPCB) 2022, the heavy metal burden in rivers is increasing, especially downstream of major industrial catchments such as Sambalpur and Cuttack, with increased lead, cadmium, and chromium contents. Furthermore, the CWC mentioned that sedimentation rates in the Hirakud dam reservoir have risen significantly in recent decades and have already reduced live storage by nearly 20%. The authors also reported that the sediment load in the river has increased because of deforestation and land cultivation upstream.

1.6.2 Impact of Dams and Water Diversion

Large dams, such as the Hirakud dam, substantially modified the natural flow of the Mahanadi River and thus changed its hydrological regime. The reduction in downstream water availability has adversely affected agriculture, fisheries, and the overall ecology of the river. The report stated that the controlled release at Hirakud dam reduces the peak discharge in the monsoon season, which, in turn, reduces natural flooding inside the deltaic area essential for maintaining the fertility of the deltaic soils (CWC 2014).

1.6.3 Impact on Biodiversity

Developmental activities in the basin have led to habitat destruction, pollution, and changes in the flow of water, all of which have had negative effects on river biodiversity. Fish species have been particularly threatened by overfishing, pollution, and migratory barriers due to dams. According to the National Biodiversity Authority (NBA, 2020), fish species diversity has declined in the Mahanadi River, with many species reaching a threatened status. The report outlined pollution, habitat fragmentation, and water diversion as the main threats to the aquatic biodiversity of the river. A study conducted by the Indian Council of Agricultural Research in 2018 on inland fisheries estimated that fish catch in the Mahanadi River has declined by 30% over the last two decades, mainly due to deterioration in riverine habitats and pollution.

1.6.4 Displacement and Livelihood Loss

Large-scale infrastructure development, such as dams and industrial complexes, has displaced entire local communities. Most of these communities have not been adequately rehabilitated, leading to continued socioeconomic nuisance. Specifically, the Ministry of Rural Development (MoRD) report of 2017 highlighted that the mere construction of the Hirakud dam caused the displacement of more than 150,000 people. Most displaced people still have difficulty accessing their livelihood, education, and health. A report titled "Displacement and Development" by NITI Aayog 2019 noted that displacement in the Mahanadi Basin has been quite high due to industry and infrastructure projects with inadequate rehabilitation, causing increased poverty and inequality in the affected regions.

1.7 Conclusion and Future Needs for Sustainable Development in the Mahanadi River Basin

The Mahanadi River Basin represents a hydrologically important entity that encompasses topographically diverse and complex landscapes in response to its rich and variable geological history, continual land use changes, and associated developmental activities. In the past two decades, rapid urbanization, deforestation, and industrialization, with the expansion of agriculture in the area, have severely affected the water quality of the basin, biodiversity, and socioeconomic conditions. All the levels of physicochemical parameters and heavy metals throughout its tributaries point to severe levels of pollution in many stretches of the river, especially in areas with industrial and urban conglomerations. Thus, holistic and integrated water management in the Mahanadi River Basin, with appropriate consideration of developmental environmental conservation, is needed. It involves mainly the promotion of pollution control measures, sustainable agricultural practices, the enhancement of waste treatment infrastructure, and the application of biodiversity conservation. Additionally, no development in sustainability can be envisaged for the region without addressing the socioeconomic challenges of displacement and inequality. With this backdrop, long-term investigations are warranted for basins in different domains, viz.:

(i) Biodiversity-rich regions: Requires multidisciplinary studies.
(ii) Industrial regions: Requires multidisciplinary studies.
(iii) River flow regimes: Environmental, ecological, and geohydrological investigations.
(iv) Urban-rurban-rural setups: Requires multidisciplinary studies.
(v) Road and railway networks: Requires pollution studies.
(vi) Water regimes: Requires multidisciplinary studies.
(vii) Command areas: Requires multidisciplinary studies with a focus on epidemiological studies.
(viii) Archaeological sites: In Chhattisgarh and Odisha, detailed investigations are needed.
(ix) Basin climate: Critical and needs attention.
(x) Resources and sustainability: Critical and needs attention.

The road map for development needs interaction between the experts from the states and formulation of a policy with holistic management models. The state governments need to prepare a database of different research outcomes of the basin and formulate a holistic management plan for sustainable development with incentivization in different sectors in coordination with different research wings within the state.

References

Ahmed SI, Sonkar AK, Kishore N, Jhariya D (2022) Evaluation of groundwater quality in Jampali coal mine, Raigarh, Chhattisgarh, India. Environ Qual Manag 31(3):9–17

Bastia F, Equeenuddin SM, Roy PD, Hernández-Mendiola E (2020) Geochemical signatures of surface sediments from the Mahanadi River basin (India): chemical weathering, provenance, and tectonic settings. Geol J 55(7):5294–5307

Bhaskar M, Dixit AK, Ojha KK, Dubey S, Singh A, Abhishek A (2020) The impact of anthropogenic organic and inorganic pollutants on the Hasdeo River water quality in Korba region, Chhattisgarh, India. Bioinformation 16(4):332

Chakrapani GJ, Subramanian V (1990) Preliminary studies on the geochemistry of the Mahanadi River basin, India. Chem Geol 81(3):241–253

Chakrapani GJ, Subramanian V (1993a) Heavy metals distribution and fractionation in sediments of the Mahanadi River basin, India. Environ Geol 22:80–87

Chakrapani GJ, Subramanian V (1993b) Rates of erosion and sedimentation in the Mahanadi River basin, India. J Hydrol 149(1–4):39–48

Chandrakar AK, Dhuria SS (2020) Sustainable wetland management in India: charting a course for a sustainable future. Life Sci Int Res J 7(1):46–52

CWC (2014). (CWC) (2014) Mahanadi Basin, Central Water Commission Version 2.0, p.109

Dadhwal VK, Aggarwal SP, Mishra N (2010) Hydrological simulation of Mahanadi River basin and impact of land use/land cover change on surface runoff using a macro scale hydrological model

Das A (2022) Surface water quality modelling using water quality index (WQI) and geographic information system (GIS) on the Mahanadi Basin, Odisha. In: International conference on water technologies. Springer Nature Singapore, Singapore, pp 21–46

Das A (2023) Modelling of surface water quality and spatial mapping in Mahanadi Basin, Odisha. J Mines Metals Fuels 71(9)

Das M, Panda T (2010) Water quality and phytoplankton population in sewage fed river of Mahanadi, Orissa, India. J Life Sci 2(2):81–85

Das A, Tripathy B (2019) Water quality monitoring and its assessment on Mahanadi basin. IOSR J Mech Civil Eng 18(3):39–55

Dhanesh S, Prasad SJ, Kumar JA (2014) Characterization of water and sediment quality of river Ahiran in Korba Chhattisgarh, India. Res J Recent Sci

Dixit PR, Kar B, Chattopadhyay P, Panda CR (2013) Seasonal variation of the physicochemical properties of water samples in Mahanadi Estuary, East Coast of India

Gardia D, Mishra AK, Khamari A (2022) A water quality analysis of the Ong River, Padampur, Bargarh, Odisha. Int J Environ Ecol Res 4:35–37

Hussain J, Dubey A, Hussain I, Arif M, Shankar A (2020) Surface water quality assessment with reference to trace metals in river Mahanadi and its tributaries, India. Appl Water Sci 10(8):1–12

India-WRIS (2014) India Water Resources Information System (n.d.). https://indiawris.gov.in/wris/#/lulc

Joy KJ, Bhadbhade N, Bhagat S, Samuel A (2021) Mahanadi Basin: An Alternative Approach to Interstate Water Sharing and Management: An approach Paper, Forum for Policy Dialogue on Water Conflicts in India, Pune, p. 43

Kar PK, Pani KR, Pattanayak SK, Sahu SK (2010) Seasonal variation in physico-chemical and microbiological parameters of Mahanadi River water in and around Hirakud, Orissa (India). Ecoscan 4(4):263–271

Mise SR, Mujawar S (2017) Evaluation of water quality of Kharun river stretch near the Raipur city. Int Res J Eng Tech 4(9):1071–1078

Mishra S, Nayak S, Pati SS, Nanda SN, Mahanty S, Behera A (2018) Spatiotemporal variation of Phytoplankton in relation to physicochemical parameters along Mahanadi estuary & inshore area of Paradeep coast, north east coast of India in Bay of Bengal

Naha S, Rico-Ramirez MA, Rosolem R (2021) Quantifying the impacts of land cover change on hydrological responses in the Mahanadi river basin in India. Hydrol Earth Syst Sci 25(12):6339–6357

Paital B, Pati SG, Panda F, Jally SK, Agrawal PK (2023) Changes in physicochemical, heavy metals and air quality linked to spot Aplocheilus panchax along Mahanadi industrial belt of India under COVID-19-induced lockdowns. Environ Geochem Health 45(3):751–770

Panigrahy BK, Raymahashay BC (2005) River water quality in weathered limestone: a case study in upper Mahanadi basin, India. J Earth Syst Sci 114:533–543

Pradhan DS, Sahu HB, Sahoo B (2017) Study of surface runoff and its impact from opencast coal mines in the IB Valley Basin, Odisha. In: International conference on deep excavation, energy resources and production

Sahoo BP, Sahu HB, Pradhan DS (2021) Hydrogeochemistry and surface water quality assessment of IB valley coalfield area, India. Appl Water Sci 11(9):153

Sarwa S, Choubey S, Thakur PK (2019) Water quality assessment of physicochemical parameters & heavy metals contamination in surface water of Jonk River, Kasdol Area, Baloda Bazar, Chhattisgarh

Sundaray SK, Nayak BB, Kanungo TK, Bhatta D (2012) Dynamics and quantification of dissolved heavy metals in the Mahanadi River estuarine system, India. Environ Monit Assess 184:1157–1179

Vaishnav MM, Hati M, Rahangdle PK (2014) Assessment of Water Pollution of Hasdeo River and its Surrounding Areas at Janjgir Champa District (C.G.) India. Int J Adv Res 2:721–727

Wani H (2008) Geochemical and petrological studies of supracrustal rocks of baster craton and its implications to crustal evolution (Doctoral dissertation). Aligarh Muslim University

A Systematic Review for Conducting a Meta-analysis: Synthesis from Multiple Studies in the Coastal Ecosystem of Odisha

2

Ashok Kumar Behera and Rachna Chandra

Abstract

The Preferred Reporting Items for Systematic Reviews and Meta-Analyses (PRISMA) is a set of rules designed to enhance the reporting and transparency of meta-analyses and systematic reviews. Following this reporting guideline, we thoroughly evaluated, synthesised, and focused on studies published on the Mahanadi marine ecosystem before and after the 1999 super cyclone in Odisha, India. The Odisha coastline provides a unique habitat with six larger estuaries, Asia's largest brackish water lagoon, the largest Olive ridley turtle population, two Horseshoe crab species, attractive tourism locations, and 258 km^2 of mangroves. In total, 1084 publications from 1970 to 2023 were obtained for the five coastal districts. Publications were arranged under 11 thematic areas, viz. biogeochemistry, flora, fauna, ecological services, microorganisms, management/threat/conservation, sea level rise/climate change/carbon sequestration/flood, geoinformatics, restoration, physiology/genetics, and social/livelihood/tourism. It was evident that the number of publications significantly increased after the 1999 super cyclone, reaching an all-time high in 2022. The highest number of publications was recorded for Puri district, which is a tourist destination and has the largest brackish water lagoon, i.e. Chilika. Meta-analyses revealed that studies on novel bioactive metabolites, drugs derived from the ocean, seaweed as a livelihood option, a decline in the number of taxonomists, mangrove resilience studies, ecological economics, and financial dimensions of blue carbon were largely scarce.

A. K. Behera · R. Chandra (✉)
Department of Environmental Sciences, Faculty of Agriculture, Sri Sri University, Cuttack, Odisha, India

J. P. Bhatt, A. K. Mishra (eds.), *Mahanadi River*,
https://doi.org/10.1007/978-981-96-1048-8_2

Keywords

After trends · Before trends · Meta-analysis · Odisha Mahanadi delta coast · PRISMA

2.1 Introduction

Odisha, located on the east coast of India with 480 km^2 coastline, stretches from the northern coast of Balasore to the southern coast of Ganjam district. The continental shelf is wider along the northern part of the Odisha coast than in the southern part (Mohanty et al. 2008). The Mahanadi delta is a significant delta on the east coast of India. The Mahanadi River drains the eastern region of the Chhattisgarh plain southeast of the state of Madhya Pradesh. It flows below Baloda Bazar into the state of Odisha and then traverses the Eastern Ghats to reach the delta plains to Naraj before emerging into the Bay of Bengal (Mahalik 1996). Approximately 9500 km^2 comprises the entire Mahanadi Delta, which is bordered to the northeast by the Brahmani River and to the southwest by Chilika Lake. This delta is a complex delta formed by the coalescence of three subdeltas formed by the Devi River in the south, the Brahmani-Baitarani Rivers in the north, and the Mahanadi River in the centre (Kumar and Bhattacharya 2004). The delta comprises parallel, complicated, and highly sedimented river deltas on Odisha's east coast (Mishra et al. 2019).

The Mahanadi delta ecosystem is essentially a depositional landform that is home to six major estuaries, the world's largest Olive ridley sea turtle rookery at Gahirmatha, India's second largest mangrove forest (Bhitarkanika Sanctuary), and Asia's largest brackish water coastal lagoon (Chilika). The region also contains extensive non-vegetated wetlands and sandy beaches rich in heavy minerals. Two species of Horseshoe crabs also find their exclusive habitat in the region. In addition to having several fishing and all-weather ports, the region also has social and economic importance for the state. In addition to foreign and domestic tourists, the region also draws avifauna from across the Himalayas and turtles and crabs that migrate for various reasons (Mishra et al. 2019).

The Mahanadi marine delta in India is extremely vulnerable to natural disasters such as storm surges, cyclones, and floods because of its unique geoclimatic conditions (Hazra et al. 2022). Since the coastal economy depends on natural resources and various land use systems, sea level rise and other climate variables significantly influence the area. These natural disasters frequently cause damage to property, human life and crops (Dube et al. 2000a). From 1891 to 2019 Odisha has experienced 98 cyclones. This is the highest number of cyclones that any other state of the east Indian coast has faced (Sahoo and Bhaskaran 2016; Sarkhel et al. 2019). One of the most disastrous cyclones experienced in the state was the 1999 super cyclone. This cyclone killed more than 10,000 people and destroyed 4.45 lakh animals, 18.43 lakh hectares of crops and 75% of the standing trees in the coastal area, affecting 1.89 crore people in the 14 districts of Odisha. The cyclone destroyed nearly 90% of the mangroves in the districts of Kendrapara and Jagatsinghpur, the *Casuarina* forest from Paradip to Konark and the coastal vegetation. Other tropical cyclones such as Phailin in 2013, Hudhud in 2014 and Fani in 2019 have affected the marine ecosystem (Fanchiotti et al. 2020).

Over the years, the coastal landscapes of Odisha have attracted considerable attention from researchers, policymakers, and conservationists, owing to their ecological significance and vulnerability to various anthropogenic and natural stressors. Since super cyclone 1999, coastal Odisha has experienced a surge in scientific interest in understanding the dynamics of coastal ecosystems. Researchers are stressing the urgent need to address environmental concerns and enhance resilience in the face of climate-related hazards (Bakshi and Panigrahi 2022). The exploration of these ecosystems has led to a myriad of studies encompassing diverse thematic areas, ranging from biogeochemistry and biodiversity assessments to investigations of the impacts of climate change and conservation efforts. This collective body of research underscores the intricate interplay between natural processes, human activities, and the resilience of coastal ecosystems in the face of evolving environmental challenges (Carrasco 2021).

The present study aims to analyse the coastal ecosystem research carried out in Odisha through the rigorous and transparent requirements provided by the Preferred Reporting Items for Systematic Reviews and Meta-Analyses (PRISMA) guidelines. This systematic review and meta-analysis synthesises existing knowledge and identifies research gaps, research trends, and emerging research priorities in coastal ecosystems. This study aims to identify future research directions, conservation strategies and sustainable management practices in Odisha's coastal landscapes. Following the PRISMA guidelines, we thoroughly evaluated and synthesised previous studies to obtain an overview of the studies published on the Mahanadi marine ecosystem before and after the 1999 super cyclone in Odisha.

2.2 Methodology

(a) The eastern coast of India is home to Odisha's rich and varied coastal ecosystem. This region includes a variety of habitats, including mangrove forests, estuaries, sandy beaches and rocky shorelines (Fig. 2.1). It is located between latitudes 17.7800° N and 22.7200° N and longitudes 81.3600° E and 87.4500° E. The shoreline ecosystem is rich in biological diversity. Mangrove forests, such as those in Bhitarkanika National Park, are pivotal, as they are home to a variety of fishes, birds and invertebrates. Estuaries play a vital role in the nesting and feeding of migratory birds and commercially valuable fishes. The olive ridley turtles visit sandy beaches for mass nesting events known as *arribadas*. Several intertidal species that have evolved, find home on the rocky coasts in Odisha. Threats to this ecosystem, however, include overfishing, pollution from industrial and agricultural runoff, habitat loss from coastal expansion, and the effects of climate change, such as rising sea levels and cyclonic storms. To preserve the biodiversity of this coastal ecosystem and the means of subsistence for

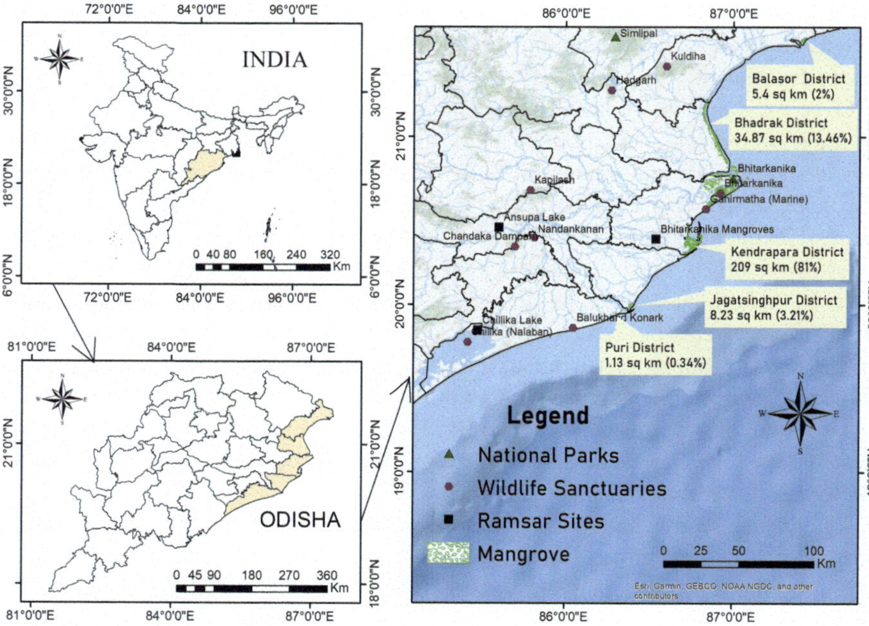

Fig. 2.1 Odisha Mahanadi marine ecosystem at different locations: Ramsar site, mangrove region, national park, wildlife sanctuary, tourist location and major rivers

the coastal communities that depend on it, conservation efforts are being made to manage it sustainably.

(b) A systematic review attempts to gather every relevant detail that satisfies specified eligibility criteria to address a particular research issue. It reduces bias in study identification, selection, synthesis and summary by using specific methodical techniques. Effective execution yields dependable results that facilitate the derivation of conclusions and the formulation of judgments (Antman et al. 1992). A systematic review must have the following essential components: (i) a clearly defined set of objectives with an explicit, repeatable methodology, (ii) a systematic search to find all the studies that fulfil the eligibility criteria, (iii) an evaluation of the validity of the included studies and (iv) a systematic presentation and synthesis of the features and findings of the included studies. Systematic reviews are frequently used in health research but are less popular in environmental studies (Greenhalgh et al. 2005; Ford et al. 2010; Moher 2019). This is in contrast to the conventional literature reviews presented in Odisha Mahanadi delta marine research, which include more narrative publications and usually do not offer specifics about the review processes utilised (such as databases, articles and search terms used). It is challenging to conduct research to verify the interpretation or assess the study's comprehensiveness due to the lack of such information. A systematic review is determined by its action, discoveries and reporting clarity.

(c) Meta-analysis, which may or may not be included in systematic reviews, is the process of combining and summarising findings via statistical methods (Moher et al. 2009).
(d) The PRISMA statement was utilised in this study, and it adopts the definition provided by the Cochrane Collaboration. 'A systematic review is an analysis of a well-defined research issue that uses specific and systematic procedures to find, evaluate and critically assess pertinent studies and gather and process data from the studies that are part of the review. The results of the included research may or may not be analysed and summarised via statistical methods (Higgins et al. 2011). The PRISMA statement is applied to help with the thorough search and classification of data in future marine environmental assessments, which are necessary in the absence of systematic reviews of English peer-reviewed publications. One can use this methodology to track advancements in studies on the marine ecosystem of the Odisha Mahanadi delta.
(e) We looked for studies assessing the relationship between January 1, 1970 and December 31, 2023, in publications related to the Odisha coastal region (Mahanadi delta). Almost 50 years of publications were used to gather these studies. We used the main databases, e.g. PubMed, Web of Science, Scopus, Google Scholar, BIOSIS Previews, Aquatic Sciences and Fisheries Abstracts (ASFA), Oceanic Abstracts, SpringerLink, ScienceDirect and ResearchGate. Only publications published in English were eligible for the search and full-text review. Research with a management study design, case series and peer-reviewed observational case reports from the Odisha Mahanadi delta coastal region were considered (Table 2.1).
(f) The literature review process used a PRISMA 2020 four-phase flow diagram (Fig. 2.2), which illustrates the systematic review process for identifying eligible studies. Initially, 1877 records were identified through sources such as Google Scholar, ResearchGate, ScienceOpen and Semantic Scholar. These records were assessed based on titles and abstracts, resulting in 1084 publications evaluated for eligibility, with 381 overlapping records. In total, 1084 records were considered for review after removing duplicates and overlapping records. An emphasis on publications produced in the coastal areas of the Mahanadi Delta in Odisha was evident in the third phase, which included full-text articles that were evaluated and included, along with the reasons for exclusion. These publications were categorised into 11 thematic areas: biogeochemistry, flora, fauna, ecological services, microorganisms, management/threat/conservation, sea level rise/climate change/carbon sequestration/flood, geoinformatics, restoration, physiology/genetics, and social/livelihood/tourism. The last step involves important studies that are a component of qualitative and quantitative synthesis (Moher et al. 2009).

Table 2.1 Eligibility criteria used in guiding the systematic literature review

Criterion	Eligibility criteria
Timeline or period	Studies published from Jan. 1, 1970 to Dec. 31, 2023
Language	Publications available in English language
Literature type	All publication in any journal with included peer-reviewed observational case reports, case series or studies
Publication status	Published and e-view (DOI—Digital object identifier) articles, articles in press and reviews
Sector	Odisha Mahanadi coastal delta ecosystem
Locations	Odisha's five major coastal districts
General topic	Biogeochemistry, flora, fauna, ecological services, microorganisms, management/threat/conservation, sea level rise/climate change/carbon sequestration/flood, geoinformatics, restoration, physiology/genetics, social/livelihood/tourism

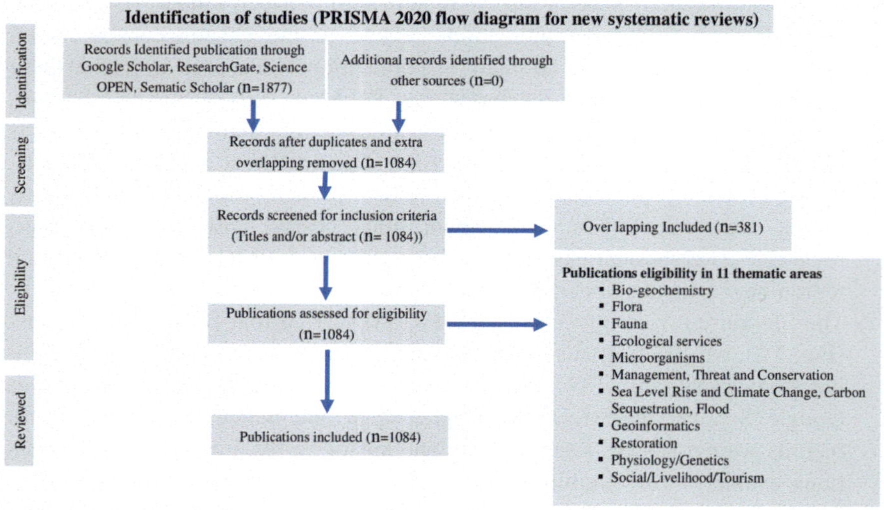

Fig. 2.2 The systematic review process included the number of papers reviewed in every step

2.3 Results and Discussion

2.3.1 About PRISMA

Systematic reviews are better than ordinary reviews because they provide the benefit of extensive literature evaluations and help identify information needs or research gaps. The approach can be utilised to keep track of publications published before and after super cyclone studies and address research gaps. The application of the PRISMA statement in maritime regions improves future review rigour, makes systematic searches easier and provides sufficient documentation of any theme.

2.3.2 Total Number of Publications on the Mahanadi Marine Ecosystem (1970–2023)

The Odisha Mahanadi delta marine region offers an interesting tourism destination and is a unique biodiversity area (Mohanty et al. 2008; Panigrahi et al. 2022). Owing to its unique geoclimatic conditions, the Mahanadi delta region is vulnerable to natural disasters such as storm surges, cyclones and floods (Dube et al. 2000a). On average, two cyclones make landfall annually in this region (Sarkhel et al. 2019). Over the last 121 years, 1301 tropical cyclonic storms have formed over the Bay of Bengal (1891–2011). Of these, 33% (389) made landfall on the coast of Odisha. One of these cyclones, the '1999 super cyclone', passed through the Mahanadi delta region, and its impact on biodiversity, tourism and livelihood was considerable (Dube et al. 2000b; Acharyya et al. 2020). Approximately 75% of standing trees, 90% of mangroves in the districts of Kendrapara and Jagatsinghpur, the *Casuarina* forest from Pradip to Konark and coastal vegetation were destroyed by the Super cyclone. Cyclones have also had damaging effects on Chilika biodiversity, coastal tourism, turtle nesting sites and the economies of coastal communities (Behera 2014; Yadav and Barve 2017; Manisha et al. 2023; Mishra et al. 2023a). The Mahanadi coastal delta region is also subject to floods, sea level rise, soil erosion and shoreline changes.

Researchers are primarily interested in quantifying the impact of a changing climate and largely focus on impacted places and other impacted biodiversity areas to analyse their findings (Frank and Marks 1998; Hossain et al. 2018; Martzikos et al. 2021; Alex et al. 2022). Researchers' interest in the Mahanadi delta region is triggered by the presence of a unique biodiversity that thrives in the region and the frequent cyclonic events that occur in the region. In the 1970s and 1980s, there were generally fewer publications (Fig. 2.3), ranging from zero to six publications per year, with occasional peaks in 1977 and 1981, with five and six publications, respectively. The 1990s marked a noticeable uptick in research, with several years seeing five to nine publications annually, including a peak of nine publications in 1999. The 2000s witnessed significant growth, with publication numbers ranging from 12 to 32 per year. Notable increases were in the year 2000 (17 publications), 2004 (21 publications), and 2005 (22 publications). Throughout the 2010s, there was continued expansion in research output, highlighted by peaks in 2010 (32 publications), 2013 (42 publications), and 2015 (44 publications), reflecting sustained interest and investment in coastal studies. This upward trend continues in the 2020s, with substantial peaks in 2022 (107 publications) and 2023 (91 publications), marking the highest levels of publications in the dataset. The publication trends in coastal research in Odisha depict a significant evolution over time. Before the 1999 super cyclone, research output from 1970 to 1998 was characterised by sporadic publication rates, ranging modestly from zero to nine publications annually. Approximately 90% of the total research papers were published after the 1999 super cyclone. Publications increased whenever a cyclone made landfall in the Mahanadi delta marine region. The cyclones, e.g. Phailin in 2013, Hudhud in 2014, Titli in 2018, Fani in 2019, Bulbul in 2019, Amphan in 2020, Yaas in 2021, Gulab in 2021 and

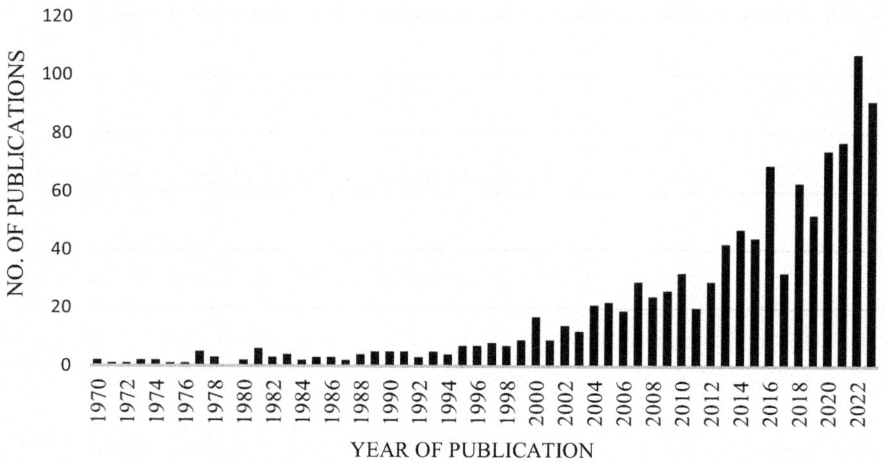

Fig. 2.3 Annual publications (1970–2023) on the Odisha Mahanadi delta marine region

Jawad in 2021, marked an increased number of scholarly publications on the Mahanadi coastal region.

2.3.3 District-Wise Publications from 1970 to 2023

Among the five coastal districts in Odisha, 29% of the total publications were from Puri, 21% from Balasore, 19% from Kendrapara, 16% from Bhadrak, and 15% from Jagatsinghpur (Fig. 2.4). A significant increase in publications from the early 2000s was recorded in Balasore. This is attributed to its rich biodiversity and vulnerability to erosion and climate change. Bhadrak, home to Bhitarkanika National Park, recorded steady growth in publications, particularly from the mid-2000s, with a peak at approximately 2020–2022, driven by its importance for wildlife conservation and climate studies. In Jagatsinghpur, a moderate increase in publications was observed from the late 2000s, and the number of publications was highest from 2018 to 2020. In Kendrapara, a significant increase in publications was recorded from the early 2010s, and the maximum number of publications occurred from 2020 to 2022. Kendrapara has drawn attention because of its mangrove ecosystem and cyclone vulnerability. Puri also recorded an increase in publications from the late 1990s onwards. Puri is rich in cultural heritage and biodiversity and faces climatic vulnerability, thus attracting the attention of many researchers. Irrespective of districts, the number of publications increased significantly in the Mahanadi delta region. The increased number of publications is driven by enhanced research funding and support, technological advancements in methodologies, increasing environmental awareness, the urgent need to address the impact of climate change, the need to document and protect biodiversity, and the focus on sustainable socioeconomic development practices such as ecotourism. Additionally, collaborations between

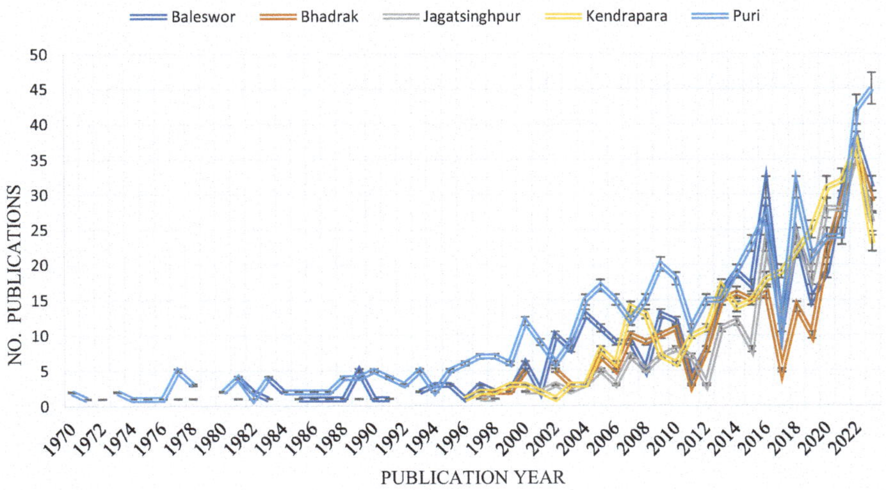

Fig. 2.4 Total number of publications per year from 1970 to 2023 on Mahanadi delta

local and international researchers also contributed to the surge in research activities and publications, providing diverse expertise and resources.

2.3.4 Thematic Area-Wise Publication

Published research on the Mahanadi delta region was organised into 11 thematic areas, i.e. (i) biogeochemistry, (ii) flora, (iii) fauna, (iv) ecological services, (v) microorganisms, (vi) management/threat/conservation, (vii) sea level rise/climate change/carbon sequestration/flood, (viii) geoinformatics, (ix) restoration, (x) physiology/genetics and (xi) social/ livelihood/tourism (Fig. 2.5). The highest number of publications (19%) were on management/threat/conservation, followed by social/ livelihood/tourism (16%). Sea level rise/climate change/carbon sequestration/floods have also been extensively studied (15%) because of their impact on ecosystems and communities. The fauna thematic area contributed 11%, as it is crucial for understanding species distributions and ecosystem health. Biogeochemistry (10%) and ecosystem services contributed to 8% of the studies, and flora studies aided in conservation efforts and contributed to 8% of the studies. Geoinformatics contributed 7% and assisted in monitoring and decision-making. The least contributed thematic areas were microorganisms with 3%, restoration with 2% and physiology/ genetics with 1%. Researchers are attracted to these areas because of their importance in managing and conserving coastal ecosystems amid growing environmental challenges.

2.4 Thematic Areas

2.4.1 Management/Threat/Conservation

With the increasing impact of climate change, coastal ecosystems throughout the world are receiving attention from researchers across disciplines. Concerns are related to coastal threats, vulnerability, management, conservation, and resilience (Tundi et al. 2011). Coastal threats are discussed in terms of overexploitation of resources, pollution, habitat degradation, rising sea levels, and climate change. Researchers are emphasising on the need to understand the risks and vulnerabilities of a region. Emphasis is placed on designing efficient management and conservation plans to protect ecosystem services, biodiversity, and means of livelihood for coastal populations (Moser et al. 2012; Senapati and Gupta 2014). Managing the rich biodiversity of the region and diverse ecological habitats has emerged as a research hotspot. Conservation advocacy aims at conservation and economic development, which ensures the sustainable utilisation of resources available to local communities (Stephania 2024). These rising concerns in the region are further complemented by international treaties and bodies, such as the Convention on Biological Diversity. Conservation goals are becoming inclusive by giving equal importance to human needs (de Oliveira Caetano et al. 2023). Researchers have also focused on the conservation of natural heritage as a process of trade-offs between socioeconomic development through tourism and the conservation of nature (Zhang et al. 2023).

Among the total publications, 19% fall under the thematic area covering management/threat/conservation (Fig. 2.5). The publications under these themes are mainly from the Puri, Kendrapara, and Balasore districts. In this context, researchers have focused on cyclones, storm surges, and risk hazards (Sahoo and Bhaskaran 2018). Conservation and management plans for the region require a thorough investigation of the risks affecting the coastal ecosystem with respect to cyclones,

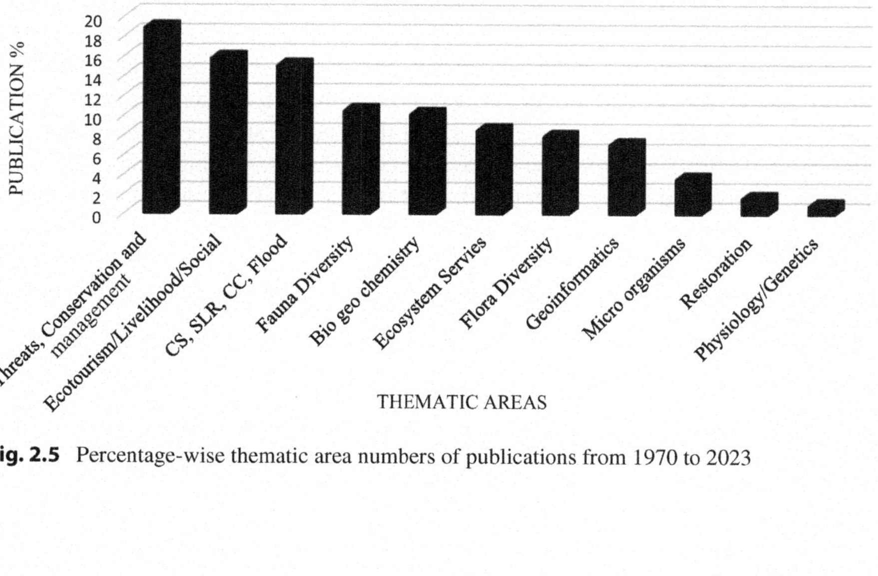

Fig. 2.5 Percentage-wise thematic area numbers of publications from 1970 to 2023

deforestation, and farming. All these factors adversely impact the biodiversity of Mahanadi delta, Bhitarkanika Sanctuary, Gahirmatha Sanctuary and Chilika Lagoon (Acharyya et al. 2020; Panigrahi et al. 2022; Mishra et al. 2023b). Studies on Chilika have focused on the changing ecological characteristics of lagoons, anthropogenic activities such as aquaculture and the natural dynamics of shifting the mouth. The conflict between livelihood and conservation interventions has been a vibrant topic of research in the social science discipline. Researchers have also investigated issues of soil erosion and land degradation in the Mahanadi delta region. Changes in the shoreline, increases in sea level and shifts in carbon potential capacity are also hotspots of research carried out in the region (Gomes et al. 2021; Mishra et al. 2022). The complex and diverse nature of the social–ecological dynamics of the Mahanadi region has become a crucial parameter for developing effective coastal management and conservation methods (Mohanty et al. 2008; Pati et al. 2022). The results of this research help shape sustainable policies that strike a compromise between the requirements of coastal populations and the preservation of important ecosystems (Pinto et al. 2014; Sandifer et al. 2015).

2.4.2 Social/Livelihood/Tourism

The Odisha region's local economy is highly dependent on natural resources such as forest products and fishing. The ecological dynamics of a region are closely linked with the socioeconomic development of the local community and beyond (Mishra 2017). The complex relationships between the livelihood and conservation of the delta and its resources have been studied. Conservation interventions in the delta region have significantly influenced community well-being (Leauthaud et al. 2013). Conservation approaches implemented in the Bhitarkanika and Gahirmatha wildlife sanctuaries are studied from the livelihood perspective of local communities, which are economically and culturally dependent on resources. These studies explore various dimensions of the complexity of socioecological relationships (Masterson et al. 2019; Karst 2021). Researchers have investigated the interlinkages among the gender division of labour, migration and environmental changes in deltaic regions. The potential of ecotourism to promote both economic development and conservation practices has been discussed by many researchers (Kiper 2013).

Puri and Balasore had the highest number of publications in this thematic area (Fig. 2.6). This may be attributed to the presence of tourist destinations in Puri. Additionally, the presence of Chilika, the largest lagoon in Asia and a Ramsar site, provides livelihoods to coastal fishermen (Mishra 2017). Recently, Chilika has emerged as a major tourist destination. The coastal belt of the Balasore district was investigated for the occurrence and management of floods, agricultural dynamics, and mangrove restoration. The promotion of ecotourism in the Bhitarkanika region of Kendrapara has also been researched for its socioeconomic and ecological impacts (Das and Chatterjee 2023). The emerging fields of investigation in the region include impact of ecotourism on indigenous communities, cultural preservation, community empowerment and the allocation of financial gains.

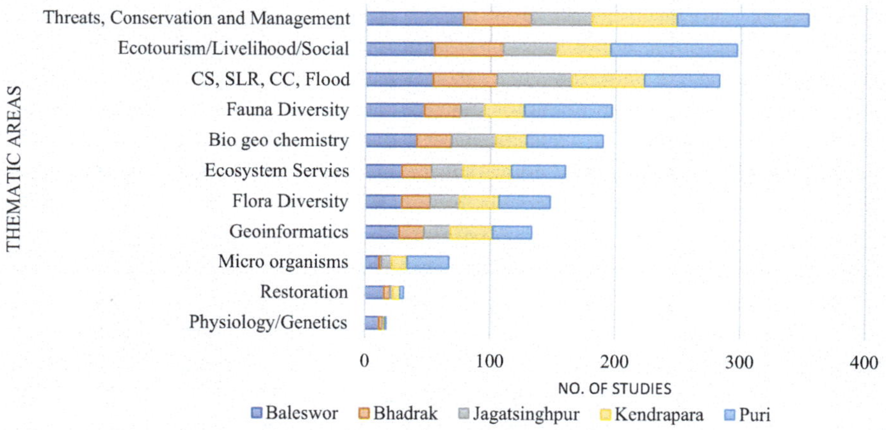

Fig. 2.6 District-wise distribution of studies in various thematic areas

2.4.3 Sea Level Rise/Climate Change/Carbon Sequestration/Flood

Sea level rise (SLR), climate change, carbon sequestration, and flood dynamics have encouraged researchers worldwide to explore these complex domains (IPCC 2019). Comprehensive research is essential for developing effective mitigation and adaptation methods because of the interdependence of various environmental elements and their effects on ecosystems, communities and global sustainability (Klein et al. 2004; Nicholls and Cazenave 2010). The characteristics of the Mahanadi delta region and its vulnerability to climate change-induced SLR, frequent occurrence of cyclic storms, floods, and denudation have been researched (Patel et al. 2019; Pradhan 2018). Odisha's extensive coastline is associated with SLR, making it imperative to understand the potential consequences for coastal communities, infrastructure, and ecosystems. Climate change intensifies these threats, affecting weather patterns, precipitation and the frequency and intensity of extreme events, including cyclones and floods (Kantamaneni et al. 2022). Researchers have highlighted the importance of these deltaic ecosystems, which are rich in mangroves, in storing carbon and contributing significantly to the world's climatic resilience (Kadaverugu et al. 2022). Flooding is another characteristic of the region that has been studied meticulously. Many dimensions of flood management interventions to suit the geophysical dynamics of the delta region have been discussed (Mohanty et al. 2020). To increase resilience and preserve the welfare of the local population, understanding how flooding, SLR and climate change interact is vital (Sharma 2019). Overall, researchers are motivated to investigate SLR, climate change, carbon sequestration and floods in Odisha to produce information that supports worldwide efforts to lessen the effects of climate change on vulnerable areas that support climate resilience and inform sustainable policies.

2.4.4 Faunal Diversity

Investigations have been carried out on coastal faunal diversity due to the ecological importance of coastal ecosystems and their ecological services (Singh et al. 2012). Coastal faunas, including fish, invertebrates, birds and other wildlife, play crucial roles in maintaining the health and functionality of coastal ecosystems (Bergstro et al. 2014). Studying coastal faunal diversity is essential for understanding complex interactions, the dynamics of marine food webs and species responses to environmental changes (Abrantes et al. 2015). These ecosystems support a rich array of marine and terrestrial species, many of which are adapted to unique conditions found along coastlines. The coastal faunas in Odisha are not only ecologically significant but also essential for the livelihoods of coastal communities, as fisheries and aquaculture are integral to the region's economy (Ailor et al. 2021). Furthermore, the vulnerability of coastal areas to climate change, pollution and habitat degradation underscores the importance of researching coastal faunal diversity in Odisha. By studying the distribution, abundance and behaviour of coastal species, researchers aim to assess the health of these ecosystems, identify potential threats and develop conservation and management strategies. Insights gained from the research on Odisha's coastal faunal diversity have contributed to the sustainable use of marine resources, biodiversity conservation and the development of policies that promote the resilience of coastal ecosystems and the communities dependent on them (Santhanam and Kundu 2022).

2.4.5 Biogeochemistry

Biogeochemistry is interdisciplinary in nature and has been extensively studied by researchers. Through these studies, the complex interactions among living organisms, geological processes and chemical cycles are explored (Ranaweera et al. 2023). By investigating the biogeochemical processes, researchers aim to understand how organisms influence the environment and vice versa and how these interactions impact ecosystem health, nutrient cycling, climate dynamics and even human livelihoods (Herndon et al. 2020).

Among the coastal districts, the maximum number of publications were from Puri (Fig. 2.6). Researchers are likely motivated by the region's unique ecological characteristics and environmental challenges (Collins et al. 2011). The state's coastal plains, forests, wetlands and mineral-rich soils provide fertile ground for studying a wide range of biogeochemical processes. Moreover, Odisha's vulnerability to natural disasters such as cyclones and floods makes it crucial to understand the biogeochemical dynamics of the region for effective environmental management and disaster mitigation strategies. It may also be necessary to address biogeochemical factors to understand and mitigate environmental problems.

2.4.6 Ecosystem Services

Ecosystem services encompass a wide range of contributions, including providing services such as food and water, regulating services such as climate and disease control, supporting services such as nutrient cycling and soil formation and providing cultural services such as recreation and spiritual well-being (Bennett et al. 2015). Understanding and quantifying these services are crucial for informed decision-making, sustainable resource management and the development of effective conservation strategies (Scholes 2017; Bullock et al. 2018). Coastal ecosystems, mangrove forests, inland forests, wetland ecosystems and agricultural areas in Odisha provide many environmental goods and services that support the livelihoods of local communities and contribute to regional and global well-being (Kadaverugu et al. 2022). Moreover, the susceptibility of coastal Odisha to natural disasters, such as cyclones and floods, underscores the importance of comprehending and valuing the ecosystem services in the region for disaster resilience and adaptive management (Bahinipati and Sahu 2012). By assessing ecosystem services, scientists aim to provide sustainable development practices, enhance resilience to environmental challenges and foster a harmonious relationship between human societies and their natural surroundings. They also aim to assess the value of natural capital and identify trade-offs and synergies between different ecosystem services (Orchard et al. 2020; Pradhan and Goswami 2024).

2.4.7 Floral Diversity

Researchers are intrigued by the study of floral diversity because of the profound impact that plant life has on the health and functioning of ecosystems and its implications for human well-being. Floral diversity plays a crucial role in ecological stability, nutrient cycling and providing habitat and food resources. Moreover, plants contribute to essential ecosystem services such as air and water purification, climate regulation and soil fertility (Ina 2016). The coastal flora, comprising the plant species found at the unique interface between land and sea, contributes significantly to shoreline stabilisation, nutrient cycling and habitat provision for numerous organisms. Investigating coastal floral diversity is crucial for understanding the resilience and adaptability of ecosystems in the face of climate change, anthropogenic impacts and SLR (Gattuso et al. 2018; Bandh et al. 2023).

Odisha's diverse climatic and topographic conditions resulted in the existence of a wide array of plant species, some of which may be unique to the region. Researchers are particularly interested in studying coastal floral diversity, such as that of mangroves, salt marshes and sandy beaches. Odisha's coastal flora also plays a crucial role in providing livelihoods for coastal communities, offering medicinal resources and contributing to the cultural identity of the region (Sahu and Panda 2021). These coastal ecosystems are vital for supporting local fisheries, protecting against coastal erosion and serving as breeding grounds for various marine species (Kumar 2021). Vulnerability to natural disasters is important in studying the floral diversity in

Odisha (Bakshi and Panigrahi 2022). The insights gained from studying Odisha's coastal floral diversity also contribute to the sustainable use of coastal resources, resilience-building in the face of climate change and the overall well-being of coastal communities (Kadaverugu et al. 2021; Barik and Ratha 2024).

2.4.8 Geoinformatics

Researchers are interested in acquiring, analysing and interpreting geographic information because of its transformative potential in addressing complex spatial challenges. Geoinformatics combines technologies such as geographic information systems (GISs), remote sensing and spatial analysis to provide valuable insights into environmental processes, resource management, disaster monitoring and urban planning (Ste 2014; Lü et al. 2019). Researchers leverage geoinformatics to optimise resource allocation, monitor ecosystem health and design sustainable development initiatives that consider the spatial dynamics of a region (Avtar et al. 2015).

Publications on geoinformatics constituted 7% of the total publications on Mahanadi delta. The Puri district constituted the maximum number of publications. Researchers have applied geoinformatics to understand various geographical and ecological characteristics and dynamics of coastal Odisha. Odisha's diverse and dynamic spatial environment is complex. Geoinformatics is a powerful tool to analyse and model diverse landscapes, enable the study of land use changes, monitor natural resources and assess the impact of climate-related events (Mishra et al. 2022; Mishra et al. 2023a). Additionally, Odisha is prone to various natural disasters, making geoinformatics an invaluable tool for disaster preparedness, response and recovery. By utilising spatial data, researchers can create accurate hazard maps, assess vulnerability and develop effective strategies for mitigating the impact of disasters on communities and infrastructure (Acharyya et al. 2020; Mishra et al. 2023b).

2.4.9 Microorganisms

Microorganisms play important roles in shaping marine ecosystems, nutrient cycling and overall environmental health. Coastal microorganisms, including bacteria, archaea and microalgae, are essential components of marine food webs and contribute significantly to the biogeochemical processes that regulate coastal ecosystems (Chandra and Kumar 2020).

In the specific context of Odisha, researchers are motivated to explore coastal microorganisms for several compelling reasons. The state's extensive coastline, encompassing diverse habitats such as mangroves, estuaries and tidal zones, creates a unique ecological setting (Chakraborty 2019). Coastal microorganisms in Odisha contribute to nutrient cycling, organic matter decomposition, and the maintenance of water quality, impacting both marine and terrestrial life (Chandra and Kumar 2020). Furthermore, the susceptibility of coastal areas to environmental changes, including SLR, climate fluctuations and anthropogenic activities, underscores the importance of studying the microorganisms in Odisha (Rahman et al. 2022). Researchers aim to understand how these microorganisms respond to environmental stressors and how their dynamics influence the resilience of coastal ecosystems. Odisha's reliance on coastal resources for fisheries, aquaculture and other livelihood activities further emphasises the importance of researching microorganisms (Subramanian et al. 2023). Microbial communities in coastal waters provide insights into the health of marine ecosystems, potential impacts on seafood safety and sustainable management of coastal resources (Parlapani 2020).

2.4.10 Restoration

Researchers are passionately engaged in exploring the process of coastal restoration because of its critical role in mitigating environmental degradation, preserving biodiversity and safeguarding the livelihoods of coastal communities (Elisa et al. 2016; Hernández-Delgado 2024). Coastal ecosystems face numerous challenges, including habitat loss, pollution and the impacts of climate change, making the study of restoration strategies crucial for the resilience and sustainability of these areas (Lu et al. 2018).

Coastal Odisha contributes to the state's economy by supporting fisheries and aquaculture and providing other ecosystem services (Ngasotter et al. 2020). Researchers are interested in developing and implementing restoration strategies tailored to Odisha's unique ecological complex (Bakshi and Panigrahi 2022). Coastal restoration efforts in the state aim to increase ecosystem resilience, protect against the impacts of climate change and contribute to the sustainable use of coastal resources. The state's extensive coastline, characterised by diverse ecosystems such as mangroves, estuaries and sandy beaches, is susceptible to natural disasters such as cyclones and floods. These events can exacerbate coastal erosion and habitat degradation, emphasising the need for effective restoration measures. Moreover, the socioeconomic importance of coastal areas in Odisha adds another layer of significance to research on coastal restoration (Mishra et al. 2022). Understanding how restoration initiatives affect local communities and their traditional practices and livelihoods is crucial for fostering community engagement and ensuring the success of restoration projects (Fox and Cundill 2018).

2.4.11 Physiology/Genetics

By analysing their physiology and genetics, researchers can comprehend the adaptation strategies of marine species and the possible implications for their conservation and sustainable resource management. Research on coastal physiology and genetics provides information on how organisms react to climate change, human influence, and environmental stressors (Rajpurohit and Jhang 2015; Muruganandam et al. 2023).

Only 1% of the publications focused on the physiology and genetics thematic area, and major studies were conducted in the Balasore district. Marine species face different issues attributed to the unique ecological conditions found in coastal regions (Mitra and Zaman 2016). Research on coastal physiology and genetics might help in conservation measures, especially for species that might be at risk from exploitation or environmental changes (Evans and Hofmann 2012). Sustainable management techniques can be guided by knowledge of the physiological responses and genetic composition of commercially significant species, protecting both the health of the marine population and the livelihood of coastal communities.

2.5 Conclusion

The study provided information about the extensive research conducted on the Mahanadi deltaic ecosystem, including those conducted after the 1999 super cyclone in Odisha. The focus on publications from 1970 to 2023 draws attention to the potential for growing research activities by incorporating more extensive geographic coverage. Surprisingly, the year 2022 generated the most number of publications, suggesting a growing focus on understanding and reducing the consequences of cyclones. Puri district had the highest rates of publication, as it is a tourist destination and is located on the bank of Chilika. Meta-analyses revealed that studies on novel bioactive metabolites, drugs from the ocean, seaweed as livelihood options, declines in the number of taxonomists, mangrove resilience, ecological economics and financial dimensions of blue carbon remain to be largely scarce. Future studies on the Mahanadi deltaic region could benefit from interdisciplinary collaboration and improved methodologies and by prioritising community-driven solutions, given its continued vulnerability to cyclones and associated hazards. By solving the identified gaps in research and expanding upon the current reservoir of knowledge, researchers can make valuable contributions to the development of more effective approaches for cyclone reduction and sustainable development in the coastal Mahanadi Delta area. The PRISMA analysis revealed both the current knowledge and potential for future research. These opportunities can be used by researchers to improve our knowledge on the effects of cyclones, strengthen community resilience in the face of climate concerns and improve emergency response systems. Researchers should actively explore partnerships, interact with others and use modern technologies such as remote sensing and geoinformatics. The topics predominantly covered general coastal ecology and biodiversity, constrained by limited infrastructure and funding. Following the 1999 super cyclone, there was a notable surge in research publications, reflecting heightened awareness and investment in understanding climate change impacts, disaster management and coastal resilience. This period saw a shift in research focus towards addressing the vulnerabilities exposed by the disaster, with increased funding, technological advancement and a stronger emphasis on policy implications and community resilience strategies shaping the trajectory of coastal research in Odisha. Researchers may strengthen the knowledge underlying evidence-based policies and practices for coastal sustainability in Odisha by utilising creative research approaches and establishing collaborative relationships and robust research activity, potentially driven by increased funding, collaboration and heightened awareness of coastal issues in recent years.

Acknowledgement The authors are thankful to the Sri Sri University, Cuttack, Odisha, India, for the logistics.

References

Abrantes G, Barnett A, Baker R (2015) Habitat-specific food webs and trophic interactions supporting coastal-dependent fishery species: an Australian case study. Rev Fish Biol Fish 25:337–363. https://doi.org/10.1007/s11160-015-9385-y

Acharyya T, Mishra M, Kar D (2020) Rapid impact assessment of extremely severe cyclonic storm Fani on morpho-dynamics & ecology of Chilika Lake, Odisha, India. J Coast Conserv 24(3). https://doi.org/10.1007/s11852-020-00754-8

Ailor FAT, Hukla ANS, Rumbic IVT (2021) Relevance of marine spatial planning in coastal zone management planning: opportunities and challenges in Indian context – case study of Odisha. J Earth Syst Sci 130(2):97. https://doi.org/10.1007/s12040-021-01574-6

Alex FJ, Tan G, Agyeman PK, Ansah PO, Olayode IO, Fayzullayevich JV, Liang S (2022) Bibliometric network analysis of trends in cyclone separator research: research gaps and future direction. Sustainability (Switzerland) 14(22):1–22. https://doi.org/10.3390/su142214753

Antman EM, Lau J, Kupelnick B, Mosteller F, Chalmers TC (1992) A comparison of results of meta-analyses of randomised control trials and recommendations of clinical experts. Treatments for myocardial infarction. JAMA 268:240–248

Avtar R, Aggarwal R, Kharrazi A, Kumar P, Agustiono T (2015) Integrating geospatial information into the implementation and monitoring of roadmaps for achieving SDGs. Environ Monit Assess 192:1–21

Bahinipati CS, Sahu NC (2012) Mangrove conservation as sustainable adaptation to cyclonic risk in Kendrapada District of Odisha, India. Asian J Environ Disast Manag 04(02):183. https://doi.org/10.3850/s1793924012001137

Bakshi A, Panigrahi AK (2022) Study on climate change and its impact on coastal habitats with special reference to ecosystem vulnerability of the Odisha coastline. In: Applied geomorphology and contemporary issues. Springer International Publishing, Cham, pp 475–497. https://doi.org/10.1007/978-3-031-04532-5_25

Bandh SA, Malla FA, Qayoom I, Mohi-ud-din H, Butt AK, Altaf A, Wani SA, Betts R, Truong TH, Dang N, Pham K, Cao DN, Ahmed SF (2023) Importance of blue carbon in mitigating climate change and plastic/microplastic pollution and promoting circular economy. Sustain For 15(3):2682

Barik A, Ratha KC (2024) Building resilience: safeguarding Odisha from natural calamities. Int J Political Sci Governance 6(1):177–183

Behera SK (2014) Poultry, fisheries & wildlife science impacts of the super cyclone Philine on sea turtle nesting habitats at the. Poult Fish Wildl Sci 2(2):1–5. https://doi.org/10.4172/pfw.1000114

Bennett EM, Cramer W, Begossi A, Cundill G, Díaz S, Egoh BN, Geijzendorffer IR, Krug CB, Lavorel S, Lazos E, Lebel L, Martín-López B, Meyfroidt P, Mooney HA, Nel JL, Pascual U, Payet K, Harguindeguy NP, Peterson GD, Woodward G (2015) Linking biodiversity, ecosystem services, and human well-being: three challenges for designing research for sustainability. Curr Opin Environ Sustain 14:76–85. https://doi.org/10.1016/j.cosust.2015.03.007

Bullock C, Joyce D, Collier M (2018) An exploration of the relationships between cultural ecosystem services, sociocultural values and well-being. Ecosyst Serv 31:142–152. https://doi.org/10.1016/j.ecoser.2018.02.020

Carrasco DLC (2021) The knowledge status of coastal and marine ecosystem services – challenges, limitations and lessons learned from the application of the ecosystem services approach in management. Front Mar Sci 8(May). https://doi.org/10.3389/fmars.2021.684770

Chakraborty SK (2019) Bioinvasion and environmental perturbation: synergistic impact on coastal–mangrove ecosystems of West Bengal, India. In: Impacts of invasive species on coastal environments: coasts in crisis, pp 171–245. https://doi.org/10.1007/978-3-319-91382-7

Chandra P, Kumar M (2020) Contribution of microbes in the renovation of wetlands. In: Restoration of wetland ecosystem: a trajectory towards a sustainable environment, pp 101–124. https://doi.org/10.1007/978-981-13-7665-8

Collins SL, Carpenter SR, Swinton SM, Orenstein DE, Childers DL, Gragson TL, Grimm NB, Grove JM, Harlan SL, Kaye JP, Knapp AK (2011) An integrated conceptual framework for long-term social–ecological research. Front Ecol Environ 9(6):351–357

Das M, Chatterjee B (2023) Ecotourism in Bhitarkanika wildlife sanctuary, India: assessment of participation, economic benefits and conservation goals. J Ecotour:1–20. https://doi.org/10.1080/14724049.2023.2282955

de Oliveira Caetano GH, Vardi R, Jarić I, Correia RA, Roll U, Veríssimo D (2023) Evaluating global interest in biodiversity and conservation. Conserv Biol 37(5):e14100

Dube SK, Chittibabu P, Rao AD, Sinha PC, Murty S (2000a) Extreme sea levels associated with severe tropical cyclones hitting Orissa coast of India. Mar Geod 2014:37–41. https://doi.org/10.1080/01490410050030652

Dube SK, Chittibabu P, Rao AD, Sinha PC, Murty TS (2000b) Sea levels and coastal inundation due to tropical cyclones in Indian coastal regions of Andhra and Orissa. Mar Geod 23(2):65–73. https://doi.org/10.1080/01490410050030643

Elisa B, Saunders MI, Abdullah S, Mills M, Beher J, Possingham HP, Mumby PJ, Lovelock CE (2016) The cost and feasibility of marine coastal restoration. Ecol Appl 26(4):1055–1074

Evans TG, Hofmann GE (2012) Defining the limits of physiological plasticity: how gene expression can assess and predict the consequences of ocean change defining the limits of physiological plasticity: how gene expression can assess and predict the consequences of ocean change. Philos Trans R Soc B Biol Sci 367(1596):1733–1745. https://doi.org/10.1098/rstb.2012.0019

Fanchiotti M, Dash J, Tompkins EL, Hutton CW (2020) The 1999 super cyclone in Odisha, India: a systematic review of documented losses. Int J Disast Risk Reduct 51(September):101790. https://doi.org/10.1016/j.ijdrr.2020.101790

Ford AC, Marwaha A, Lim A, Moayyedi P (2010) Systematic review and meta-analysis of the prevalence of irritable bowel syndrome in individuals with dyspepsia. Clin Gastroenterol Hepatol 8(5):401–409. https://doi.org/10.1016/j.cgh.2009.07.020

Fox H, Cundill G (2018) Towards increased community-engaged ecological restoration: a review of current practice and future directions. Ecol Restor 36(3):208–218

Frank D, Marks LKS (1998) Landfalling tropical cyclones: forecast problems and associated research opportunities. Am Meteorol Soc Bull 79(2):305–323

Gattuso JP, Magnan AK, Bopp L, Cheung WW, Duarte CM, Hinkel J, Rau GH (2018) Ocean solutions to address climate change and its effects on marine ecosystems. Front Mar Sci 5:337

Gomes LE, Sanders CJ, Nobrega GN, Vescovi LC, Queiroz HM, Kauffman JB, Ferreira TO, Bernardino AF (2021) Ecosystem carbon losses following a climate-induced mangrove mortality in Brazil. J Environ Manag 297(July):113381. https://doi.org/10.1016/j.jenvman.2021.113381

Greenhalgh T, Robert G, MacFarlane F, Bate P, Kyriakidou O, Peacock R (2005) Storylines of research in diffusion of innovation: a meta-narrative approach to systematic review. Soc Sci Med 61(2):417–430. https://doi.org/10.1016/j.socscimed.2004.12.001

Hazra S, Ghosh A, Ghosh S, Pal I, Ghosh T (2022) Assessing coastal vulnerability and governance in Mahanadi Delta, Odisha, India. Prog Disast Sci 14:100223. https://doi.org/10.1016/j.pdisas.2022.100223

Hernández-Delgado EA (2024) Coastal restoration challenges and strategies for Small Island developing states in the face of sea level rise and climate change. Coasts 4(2):235–286. https://doi.org/10.3390/coasts4020014

Herndon E, Kinsman-Costello L, Godsey S (2020) Biogeochemical cycling of redox-sensitive elements in permafrost-affected ecosystems. In: Biogeochemical cycles: ecological drivers and environmental impact, pp 245–265

Higgins JPT, Altman DG, Gøtzsche PC, Jüni P, Moher D, Oxman AD, Savović J, Schulz KF, Weeks L, Sterne JAC (2011) The Cochrane Collaboration's tool for assessing risk of bias in randomised trials. BMJ (Online) 343(7829):1–9. https://doi.org/10.1136/bmj.d5928

Hossain MS, Pogue SJ, Trenchard L, Van Oudenhoven APE, Washbourne CL, Muiruri EW, Tomczyk AM, García-Llorente M, Hale R, Hevia V, Adams T, Tavallali L, De Bell S, Pye M, Resende F (2018) Identifying future research directions for biodiversity, ecosystem services and sustainability: perspectives from early-career researchers. Int J Sust Dev World 25(3):249–261. https://doi.org/10.1080/13504509.2017.1361480

Ina S (2016) Toward liveable and healthy urban streets: roadside vegetation provides ecosystem services where people live and move Environmental Science & Policy toward liveable and healthy urban streets: roadside vegetation provides ecosystem services were people li. Environ Sci Policy 62(June):24–33. https://doi.org/10.1016/j.envsci.2015.11.012

IPCC (2019) Special report on the ocean and cryosphere in a changing climate

Kadaverugu R, Dhyani S, Dasgupta R, Kumar P, Hashimoto S, Pujari P (2021) Multiple values of Bhitarkanika mangroves for human well-being: synthesis of contemporary scientific knowledge for mainstreaming ecosystem services in policy planning. J Coast Conserv 25:1–15

Kadaverugu R, Dhyani S, Purohit V, Dasgupta R, Kumar P, Hashimoto S, Pujari P, Biniwale R (2022) Scenario-based quantification of land-use changes and its impacts on ecosystem services: a case of Bhitarkanika mangrove area, Odisha, India. J Coast Conserv 26(4):1–19. https://doi.org/10.1007/s11852-022-00877-0

Kantamaneni K, Panneer S, Krishnan A, Shekhar S, Bhat LRAK, Rice L (2022) Appraisal of climate change and cyclone trends in Indian coastal states: a systematic approach towards climate action. Arab J Geosci 15(9). https://doi.org/10.1007/s12517-022-10076-8

Karst HE (2021) Social-ecological wellbeing of communities engaged in ecotourism: perspectives from Sakteng wildlife sanctuary, Bhutan social-ecological wellbeing of communities engaged in ecotourism: perspectives from Sakteng wildlife. J Sustain Tour:1–23. https://doi.org/10.1080/09669582.2021.1913500

Kiper T (2013) Role of ecotourism in sustainable development

Klein RJT, Nicholls RJ, Thomalla F (2004) Resilience to natural hazards: how useful is this concept? Global Environ Change B Environ Hazards 5(2003):35–45. https://doi.org/10.1016/j.hazards.2004.02.001

Kumar S (2021) How to fish combining indigenous, traditional and scientific fishery advisories to preserve marine provisioning ecosystem services of Odisha, India. In: Traditional and scientific fishery advisories to preserve marine provisioning ecosystem Services of Odisha, India

Kumar KV, Bhattacharya A (2004) Remote sensing of delta progradation in Mahanadi Delta, Orissa. J Geol Soc India 64(2):227–230

Leauthaud C, Duvail S, Hamerlynck O, Paul JL, Cochet H, Nyunja J, Albergel J, Grünberger O (2013) Floods and livelihoods: the impact of changing water resources on wetland agro-ecological production systems in the Tana River Delta. Kenya Global Environ Change 23(1):252–263. https://doi.org/10.1016/j.gloenvcha.2012.09.003

Lu Y, Yuan J, Lu X, Su C, Zhang Y, Wang C, Cao X, Li Q, Su J, Ittekkot V, Garbutt RA, Bush S, Fletcher S, Wagey T, Kachur A, Sweijd N (2018) Major threats of pollution and climate change to global coastal ecosystems and enhanced management for sustainability. Environ Pollut 239:670–680. https://doi.org/10.1016/j.envpol.2018.04.016

Lü G, Batty M, Strobl J, Lin H, Zhu A, Chen M (2019) Reflections and speculations on the progress in geographic information systems (GIS): a geographic perspective. Int J Geogr Inf Sci 33(2):346–367. https://doi.org/10.1080/13658816.2018.1533136

Mahalik N (1996) Geomorphology and evolution of the Mahanadi Delta, India. Oceanogr Lit Rev 4(44):323

Manisha M, Verma K, Ramesh N, Anirudha T, Santrupt R, Das RM, Kumar M, Chanakya H, Rao L (2023) Socioeconomic impact assessment of large-scale recycling of treated municipal wastewater for indirect groundwater recharge. Sci Total Environ 859:160207. https://doi.org/10.1016/j.scitotenv.2022.160207

Martzikos NT, Prinos PE, Memos CD, Tsoukala VK (2021) Key research issues of coastal storm analysis. Ocean Coast Manag 199(September):105389. https://doi.org/10.1016/j.ocecoaman.2020.105389

Masterson VA, Vetter S, Chaigneau T, Daw TM, Selomane O, Hamann M, Wong GY, Mellegård V, Cocks M, Tengö M (2019) Revisiting the relationships between human well-being and ecosystems in dynamic social-ecological systems: implications for stewardship and development. Global Sustaina 2:e8

Mishra PK (2017) Socioeconomic impacts of climate change in Odisha: issues, challenges and policy options. J Clim Change 3(1):93–107. https://doi.org/10.3233/jcc-170009

Mishra SP, Sethi KB, Barik KK (2019) Delta partitioning, geospatial changes, anastomosis of Mahanadi tri-delta, India. Int J Earth Sci Eng Caffetinnova Org 12(1):21–40. https://doi.org/10.21276/ijee.2019.12.0103

Mishra M, Acharyya T, Chand P, Santos CAG, Kar D, Das PP, Pattnaik N, Silva RM, Nascimento TVM (2022) Analysing shoreline dynamicity and the associated socioecological risk along the southern Odisha coast of India using remote sensing-based and statistical approaches. Geocarto Int 37(14):3991–4027. https://doi.org/10.1080/10106049.2021.1882005

Mishra M, Acharyya T, Santos CAG, Silva RM, Kar PK, Mohanty PK, Rout NR, Beja SK, Bhattacharyya D, Behera B, Barik S, Mahapatra S (2023a) Impact assessment of severe cyclonic storm Asani on the nesting grounds of olive Ridley turtle, Rushikulya estuary and spit in Odisha state, India. Ocean Coast Manag 238(March). https://doi.org/10.1016/j.ocecoaman.2023.106572

Mishra M, Panda GK, Dandapat K, Chatterjee U (2023b) Vulnerability assessment of the inherent hazards of climate change on the coastal environment of the Mahanadi Delta, East Coast of India. In: Ecological footprints of climate change: adaptive approaches and sustainability. Springer International Publishing, Cham, pp 157–178

Mitra A, Zaman S (2016) Basics of marine and estuarine ecology. Springer. https://doi.org/10.1007/978-81-322-2707-6

Mohanty PK, Panda US, Pal SR, Mishra P (2008) Monitoring and management of environmental changes along the Orissa Coast. J Coast Res 24(2 Suppl. B):13–27. https://doi.org/10.2112/04-0255.1

Mohanty MP, Mudgil S, Karmakar S (2020) Flood management in India: a focussed review on the current status and future challenges. Int J Disast Risk Reduct 49(May):101660. https://doi.org/10.1016/j.ijdrr.2020.101660

Moher D (2019) Preferred reporting items for systematic review and meta-analysis protocols (prisma-p) 2015 statement. Jpn Pharmacol Ther 47(8):1177–1185

Moher D, Liberati A, Tetzlaff J, Altman DG (2009) Preferred reporting items for systematic reviews and meta-analyses: the PRISMA statement. PLoS Med 6:e1000097

Moser SC, Jeffress WS, Boesch DF (2012) Wicked challenges at Land's end: managing coastal vulnerability under climate change. Annu Rev Environ Resour 37:51–78. https://doi.org/10.1146/annurev-environ-021611-135158

Muruganandam M, Rajamanickam S, Sivarethinamohan S, Reddy MK, Velusamy P, Gomathi R, Ravindiran G, Gurugubelli TR, Munisamy SK (2023) Impact of climate change and anthropogenic activities on aquatic ecosystem–a review. Environ Res:117233

Ngasotter S, Panda SP, Mohanty U, Akter S, Mukherjee S, Waikhom D, Devi LS (2020) Current scenario of fisheries and aquaculture in India with special reference to Odisha: a review on its status, issues and prospects for sustainable development. Int J Bioresource Stress Manag 11(4):370–380

Nicholls RJ, Cazenave A (2010) Sea-level rise and its impact on coastal zones. Science 328(5985):1517–1520

Orchard S, Glover D, Thapa Karki S, Ayele S, Sen D, Rathod R, Rowhani P (2020) Exploring synergies and trade-offs among the sustainable development goals: collective action and adaptive capacity in marginal mountainous areas of India. Sustain Sci 15(6):1665–1681. https://doi.org/10.1007/s11625-019-00768-8

Panigrahi AK, Chaudhury SK, Sarkar S, Gebrie BD (2022) Growth, prospects and challenges of tourism industry in Odisha. J Manag Res Analy 9(2):70–77. https://doi.org/10.18231/j.jmra.2022.015

Parlapani FF (2020) Microbial diversity of seafood. Curr Opin Food Sci. https://doi.org/10.1016/j.cofs.2020.09.005

Patel SK, Mathew B, Nanda A, Pati S, Nayak H (2019) A review on extreme weather events and livelihood in Odisha. India Mausam 70(3):551–560. https://doi.org/10.54302/mausam.v70i3.258

Pati P, Desai N, Mankad A, Solanki H (2022) The decadal changes and socio-economic stresses on coastal area through remote sensing and Gis technology: a review of Puri, Odisha. Int Assoc Biolo Comput Digest 1(2):45–53. https://doi.org/10.56588/iabcd.v1i2.41

Pinto R, De JVN, Marques JC (2014) Linking biodiversity indicators, ecosystem functioning, provision of services and human well-being in estuarine systems: application of a conceptual framework. Ecol Indic 36:644–655. https://doi.org/10.1016/j.ecolind.2013.09.015

Pradhan S (2018) Understanding the role of politics of scale and power relations in local governance of climate change adaptation: case study from coastal Odisha, India Thesis submitted for the Degree of Doctor of Philosophy. PhD Thesis, May

Pradhan RS, Goswami A (2024) Sustainability and sustainable development goals: a select study of the traditions and rituals of Odisha. In: International conference on cultural studies (ICCUS 2023). Atlantis Press, pp 200–216

Rahman S, Ahmed F, Islam S (2022) A review of the physicochemical features and phytoplankton community of the Bay of Bengal: Bangladesh perspective, p 1–18

Rajpurohit D, Jhang T (2015) In situ and ex situ conservation of plant genetic resources and traditional knowledge. In: Plant genetic resources and traditional knowledge for food security, pp 137–162

Ranaweera D, Kumarasinghe U, Senanayake C, Dassanayake RS, Bandara PC, Koliyabandara PA (2023) Biogeochemistry: essential link between geosphere and biosphere. In: Medical geology: en route to one health, pp 37–54

Sahoo B, Bhaskaran PK (2016) Assessment on historical cyclone tracks in the bay of Bengal, east coast of India. Int J Climatol 36(1):95–109. https://doi.org/10.1002/joc.4331

Sahoo B, Bhaskaran PK (2018) Multihazard risk assessment of coastal vulnerability from tropical cyclones – a GIS based approach for the Odisha coast. J Environ Manag 206:1166–1178. https://doi.org/10.1016/j.jenvman.2017.10.075

Sahu SK, Panda T (2021) Patterns of plant use in religious offerings of Odisha. Eur J Biol Res 11(3):307–314

Sandifer PA, Sutton GAE, Ward BP (2015) Exploring connections among nature, biodiversity, ecosystem services, and human health and well-being: opportunities to enhance health and biodiversity conservation. Ecosyst Serv 12:1–15. https://doi.org/10.1016/j.ecoser.2014.12.007

Santhanam H, Kundu SK (2022) Assessment of sociotechnical constraints of marine 968 fishers in the utilisation of marine fishery advisories in southern Odisha, India. Anthr Sci. https://doi.org/10.1007/s44177-022-00014-4

Sarkhel P, Biswas D, Sarita S (2019) A review of cyclone and its impact on the coastal belts of Odisha. Int J Eng Res Technol 8(5):759–762

Scholes MWRJ (2017) The GEO handbook on biodiversity observation networks. Afr J Range Forage Sci 34(4). https://doi.org/10.2989/10220119.2017.1387815

Bergstro U, Seitz RD, Lipcius RN, Ysebaert T (2014) Ecological value of coastal habitats for commercially and ecologically important species. ICES J Mar Sci 71(3):648–665

Senapati S, Gupta V (2014) Climate change and coastal ecosystem in India: issues in perspectives. Int J Environ Sci 5(3):530–543. https://doi.org/10.6088/ijes.2014050100047

Sharma V (2019) Access for adaptation? Reviewing the linkages between energy, disasters, and development in India. Energy Res Soc Sci 52:10–19. https://doi.org/10.1016/j.erss.2019.01.025

Singh AK, Ansari A, Kumar D, Sarkar UK (2012) Status, biodiversity and distribution of mangroves in India: an overview. In: Uttar Pradhesh sate biodiversity board. Marine biodiversity: one ocean, many worlds of life, pp 59–67

Ste R (2014) Geographic information science I: why does a smart city need to be spatially enabled? Prog Hum Geogr 38(5):703–711. https://doi.org/10.1177/0309132513517365

Stephania R (2024) Promotion of the conservation of biological diversity of interest for tourism in the coastal region of Ecuador. South Florida J Dev 5:516–524. https://doi.org/10.46932/sfjdv5n2-008

Subramanian A, Nagarajan AM, Vinod S, Chakraborty S, Sivagami K, Theodore T, Sathyanarayanan SS, Tamizhdurai P, Mangesh VL (2023) Long-term impacts of climate change on coastal and transitional eco-systems in India: an overview of its current status, future projections, solutions. RSC Adv 13(18):12204–12228. https://doi.org/10.1039/d2ra07448f

Tundi A, John D, Kristin SOV (2011) taking steps toward marine and coastal management

Yadav DK, Barve A (2017) Analysis of socioeconomic vulnerability for cyclone-affected communities in coastal Odisha, India. Int J Disast Risk Reduct 22:387–396. https://doi.org/10.1016/j.ijdrr.2017.02.003

Zhang Z, Xiong K, Huang D (2023) Natural world heritage conservation and tourism: a review. Herit Sci:1–15. https://doi.org/10.1186/s40494-023-00896-6

Development of an Ecologically Sensitive Prototype Plan for Riverine Island of Mahanadi at Cuttack: A Comprehensive Assessment

3

Rishab Ray and Sonal Tiwari

Abstract

To address the alarming degradation of riverbanks and their ecosystems, many studies have identified its root causes. However, each river has its own ecosystem. Therefore, this research seeks to formulate an ecologically sensitive prototype plan for the riverine islands of Mahanadi in Cuttack, with a primary focus on sustainable development and conservation. In this study, a map of the Mahanadi Basin around Cuttack city from Chahata Park to Jobra Barrage was rasterized with geographic information system (GIS), which included a detailed survey of its ecological and cultural heritage. Through a combination of field surveys, data analysis, and stakeholder consultations, this research identified key factors contributing to the environmental stressors affecting the river. Using advanced mapping techniques, this research aims to inform the development of targeted strategies for the protection of riverine islands and to create a comprehensive understanding of the complex relationship between a watershed and its cultural significance. By cataloguing and analyzing rich biodiversity, targeted strategies for the protection of riverine islands can be developed. This study further concludes with a proposal for ecologically sensitive development of riverine island guided by best practices.

R. Ray (✉)
Faculty of Architecture, Sri Sri University, Cuttack, Odisha, India
e-mail: rishab.r@srisriuniversity.edu.in

S. Tiwari
School of Planning and Architecture, Bhopal, Madhya Pradesh, India
e-mail: sonal@spabhopal.ac.in

J. P. Bhatt, A. K. Mishra (eds.), *Mahanadi River*,
https://doi.org/10.1007/978-981-96-1048-8_3

Keywords

Riverine island · Island ecosystem planning · Sustainable tourism · Environmental conservation · Mahanadi River · Riverfront development

3.1 Introduction

There has been increasing interest in riverfront development and planning. Few projects have demonstrated ecologically sensitive sustainable development on riverfronts or islands (Harvey 2016; Kumar and Sharma 2020). Concretization of edges is commonly observed on riverbanks in the name of riverbank development, which further deteriorates riparian conditions and river health (Tiwari et al. 2023). The provision of an ecologically healthy ecosystem is equally necessary in the aesthetic planning of banks.

The landscape diversity of a healthy river is often disregarded in riverfront development projects (Dournel 2024). Landscape diversity can have a major impact on the nature of rivers. While rivers usually have a well-needed bank separating channels from floodplains or valley sides, they should not be considered in isolation from meadows, elds, woods, settlements, and people on their banks. Modern scientific studies have endeavored to examine rivers from their source to the sea, but investigations of rivers can always prove challenging.

Riverine island and its ecology: An island within a river course can be called an eyot. Riverine islands are surrounded by fresh riverine water, which functions as a barrier to terrestrial animal and plant dispersion. These characteristics demonstrate ecological isolation, unique biodiversity, and endemism. The size, distance, and period of isolation from massive landmasses regularly culminate in high degrees of adaptive specialization and, for this reason, high stages of endemism. Isolation, as a spinoff of biogeography insulation, is a key factor of evolutionary change because it allows the genetic reservoir of a populace to emerge as distinct from that of other populations (Costanzi and Steifetten 2019). Island isolation has commonly led to a high degree of flora and fauna specialization related to excessive endemism, and this is mainly true for small isolated oceanic islands. Riverine islands have been discussed much less than oceanic islands. In India, all major rivers have these island ecosystems, which are in a highly degraded state due to erosion and other factors.

The second SUSE (Symposium on Urbanization and Stream Ecology) led to the identification of 26 research questions on riparian urbanization along with the modification of the previous model on stream impacts. The third SUSE emphasized that the role of humans in the modification of riverine systems must be studied. Compared with other landscapes, the riparian landscapes of these spaces harbor unique and diverse vegetation.

Diverse human chronological approaches emphasize the importance of the following: (1) assessing biophysical resources and cultural values through various sociocultural indicators using various toolsets, such as questionnaires, index-based assessments, strategic matrices, carrying capacity, impact assessments, and ecosystems (Mishra 2019), which can be utilized on the basis of their suitability; (2) holistic models that can assess the physical and nonphysical values of landscapes that are useful and can be developed through the integration of the abovementioned approaches to assess the values of landscapes and their linkages with communities.

3.2 Materials and Methodology

This study assesses the ecosystem and landscape layers of Mahanadi islands via a layer-cake model, as suggested by Ian McHarg. Ecological planning principles derived from the research are further applied (Rath 2018). Three steps were predominantly applied:

1. Assessment of natural resources

Mapping of resources such as geology, soil, vegetation on site, and wildlife was performed via standard mapping techniques and secondary data, as shown in Table 3.1.

2. Mapping of cultural resources

A detailed mapping of land cover and activities on the edges of the Mahanadi riverfront was performed. An assessment of the changing landscape was also performed.

3. Identification of critically sensitive open spaces and application of ecological principles for future development

Mapping of the various resources, as discussed in the human ecological approach, was performed to identify the ecologically fragile zone from the developable zone.

Study Area Mahanadi Riverine Island in Cuttack is a habitat for pertinent local and regional resources that influence the region's socioeconomic and environmental conditions. The island is located at the heart of Cuttack's deltaic plains, which have also historically been a hub for trade and culture. In addition to enhancing the

Table 3.1 Methods of mapping

S. no	Natural resource	Method of mapping
1	Soil and geology	Field survey, secondary data from maps of the Department of Soil Sciences and Geological Survey of India
2	Relief	Field survey, mapping on GIS platform
3	Hydrology	Field survey, mapping on GIS platform
4	Vegetation and wildlife	Field survey, quadrat study, site observations

island's natural diversity, this area helps with a number of human endeavors, including fishing, tourism, and agriculture.

3.3 Natural Resource Assessment

Geology Island topography is formed by fluvial processes related to the Mahanadi waterway. The sedimentary layers, essentially composed of alluvial deposits, make a prolific substrate that underpins rich vegetation and assorted environments. Moreover, geographical highlights incorporate common levees and floodplains, which are imperative for relieving surge dangers and supporting horticulture (Tiwari et al. 2023). These arrangements are continuously affected by the energetic stream, contributing to the island's advancing scene (Fig. 3.1).

Soil The Mahanadi River continuously deposits nutrient-rich silt onto the island, resulting in mostly alluvial soil with high fertility. Because of the exceptional suitability of this soil type for growing a wide range of crops, agriculture is the main source of income for the local population. Alluvial soils have a high ability to retain water, which, in turn, enhances their fertility and allows for the growth of both seasonal and permanent crops. Nevertheless, to maintain agricultural output and stop

Fig. 3.1 Riverine islands in Cuttack

land degradation, these soils also need to be managed well because they are prone to erosion (National Bureau of Soil Survey and Land Use Planning 2015).

Hydrology A number of small-order streams are present on the island (Fig. 3.2). During peak inundation, a significant portion of the island is flooded, leading to soil loss in these zones.

Vegetation Native grasses, shrubs, and trees make up the varied vegetation of Mahanadi Riverine Island (Figs. 3.3 and 3.4). The dense growth of aquatic and semiaquatic plants defines riparian zones, and these plants are essential for preventing erosion and stabilizing riverbanks. The island's central regions are frequently covered in deciduous forests that are home to species such as bamboo, various fruit-bearing trees, and Sal (*Shorea robusta*). In addition to serving as a habitat for wildlife, these forests give nearby communities access to resources such as fuelwood, lumber, and nontimber forest products (Rath 2018).

Wildlife Island biodiversity is considered a hotspot because of its diverse habitats, which are home to a wide variety of species (Kumari 2022; Das 2021). A multitude of animal species, including several endangered and migratory bird species, find niches in the mosaic of wetlands, grasslands, and forests. The wetlands on the island draw a wide range of waders and waterfowl, which makes it a valuable location for ornithological research and birdwatching. Furthermore,

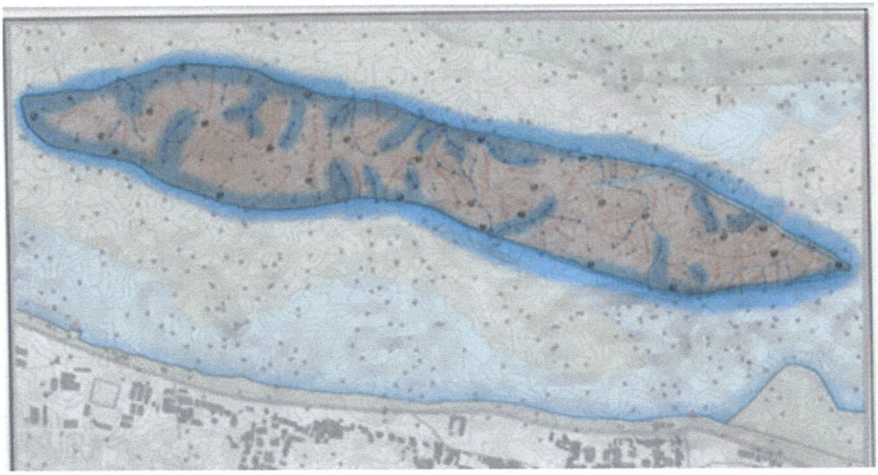

Fig. 3.2 Hydrology of island ecosystems indicating peak flood zones

Fig. 3.3 Vegetation typology on the island

Fig. 3.4 Vegetation diversity on the island

mammals such as deer, wild boar, and other small mammals are supported by terrestrial ecosystems, in addition to reptiles and amphibians, which flourish in moist, fertile environments.

Tourism The biodiversity, cultural legacy, and natural beauty of the Mahanadi Riverine Island are its main drivers of tourism (Rout 2020). Nature lovers and eco-tourists are drawn to the island's picturesque landscapes, which include views of a tranquil river and abundant vegetation. Deeply ingrained in the community's customs, cultural events and festivals provide tourists with one-of-a-kind experiences that foster cultural tourism. In addition, researchers and nature enthusiasts are drawn to the island by its wildlife and birdwatching opportunities. To prevent tourism development from jeopardizing the island's ecological integrity, sustainable tourism practices are crucial for striking a balance between the influx of visitors and the preservation of natural and cultural resources.

Cultural Services Forts, temples, traditional ghats, and significant trees represent (Figs. 3.5 and 3.6) the riparian zone as a representative of the bygone era and are examples of the connection of rivers to humans (Dournel 2024). Rituals, chants, songs, and sounds center around rivers. The rivers also spiritually play a role in daily life. The various rituals of the Hindu religion related to birth, marriage, and death are closely linked to water bodies. A comprehensive open space network can integrate heritage structures on edges with riverine elements.

3.4 Issues Identified

(i) Seasonal flooding and drought: The place is still surrounded by the subsidizing of soil. The precipitation in monsoons can occur during the hurricane season because of the main flood problem. The wastewater from the varied development on the North Bank and South Bank needs to be mitigated by strategies such as constructed wetlands and sewage treatment plants.
(ii) Soil erosion and land degradation: Soil erosion and degradation were observed on both riverbanks and islands.
(iii) Invasive plant species and loss of biodiversity: The occurrence of invasive species on islands requires systematic removal and reestablishment of native flora.
(iv) Grazing: Grazing by cattle has degraded the vegetation of the island.

3.5 Results

3.5.1 Application of Ecological Planning Guidelines

The following planning principles were adopted to develop an ecologically resilient experiential tourism plan:

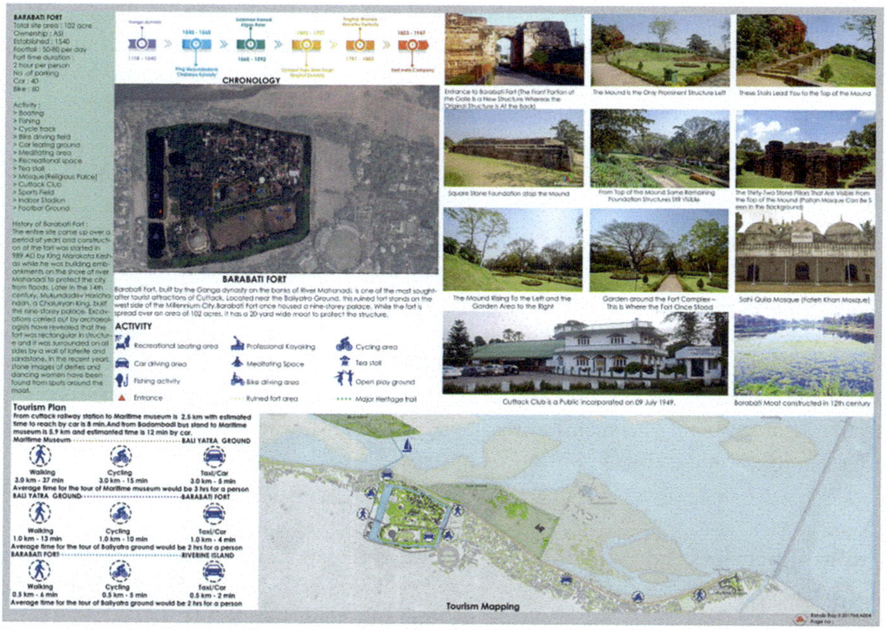

Fig. 3.5 Places of social significance on the banks of the Mahanadi River

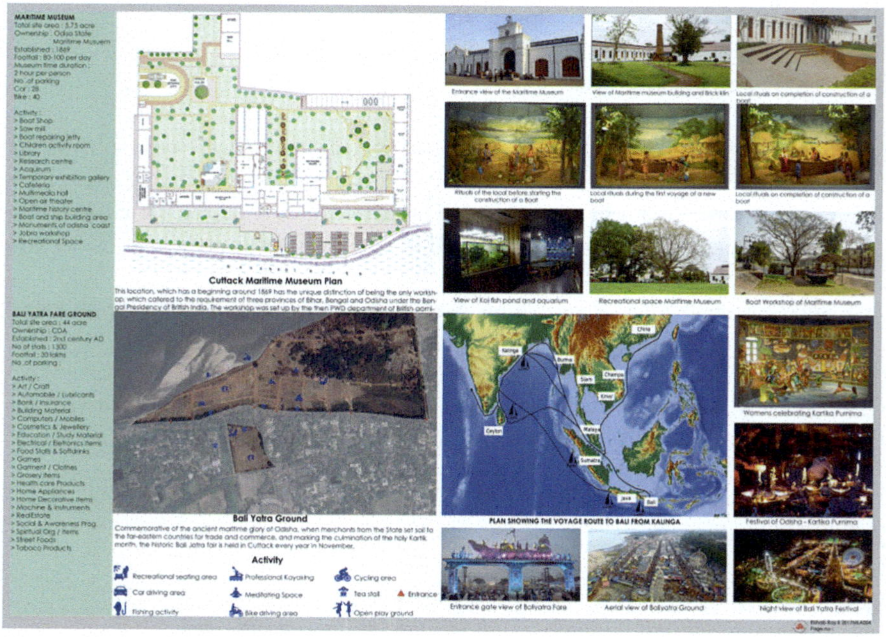

Fig. 3.6 Bali Jatra ground, whose historical significance lies in a degraded state

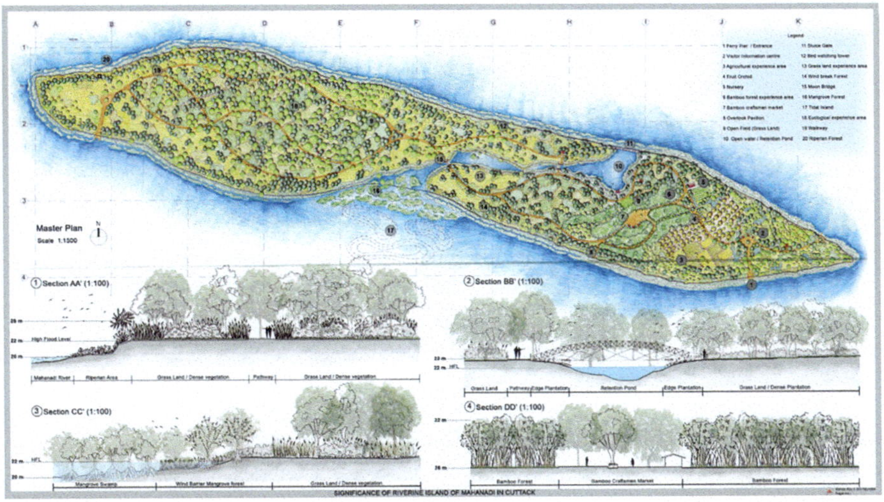

Figs. 3.7 Ecological development plan of the island with elements of passive recreation, such as earthen trails, boardwalks, bamboo view decks, and traditional boat jetty

1. Leave room for water in flood areas, and provide measures for soil conservation by providing adequate riparian zones (Singh 2018).
2. Protect the riparian vegetation zone through measures such as fencing from grazing, encroachment or illegal felling, and the influence of riparian vegetation on hydrological processes involving runoff control, water uptake by the root system, and water quality improvisation.
3. Identify the critical habitat, and conserve it within a buffer zone while respecting the current planting pattern.
4. Identify the cultural and heritage zones, and integrate them into the tourism circuit.
5. Study the issue of wastewater through decentralized systems such as decentralized wastewater treatment system (DEWATS) and constructed wetlands.
6. Identify activities for passive recreation and minimal ecological impact, and use the remaining area for ecotourism activities.
7. Provide nature trails to experience rich vegetation resource islands. Using the abovementioned principle, a comprehensive development plan is proposed. In the initial phase of 7 years, ecological restoration is proposed, followed by tourism development (Figs. 3.7 and 3.8).

3.6 Conclusion

Riverine islands are ecologically sensitive zones and thus should be conserved. The restoration of island ecology needs to be the foremost step in reintegrating these spaces. If protected, the endemic flora and fauna of the riverine island can lead to a rich biodiversity in the region. Nature-based solutions such as riparian buffer

Figs. 3.8 Ecological development plan of the island with elements of passive recreation, such as earthen trails, boardwalks, bamboo view decks, and traditional boat jetty

plantations and constructed wetlands are cost-effective solutions and provide ecological richness in the areas of intervention.

References

Costanzi JM, Steifetten Ø (2019) Island biogeography theory explains the genetic diversity of a fragmented rock ptarmigan (Lagopus muta) population. Ecol Evol 9(7):3837–3849. https://doi.org/10.1002/ece3.5007

Das S (2021) Biodiversity assessment of Mahanadi River Basin. Biodivers Conserv 30(8):2345–2360. https://doi.org/10.1007/s10531-021-02197-5

Dournel S (2024) The challenges of urban requalification of rivers and wetlands. the case of Peri-Francilian Towns. In: Farguell Pérez J, Santasusagna Riu A (eds) Urban and metropolitan rivers. The urban book series. Springer, Cham. https://doi.org/10.1007/978-3-031-62641-8_10

Harvey JW (2016) Chapter 1 – hydrologic exchange flows and their ecological consequences in river corridors. In: Jones JB, Stanley EH (eds) Stream ecosystems in a changing environment. Academic Press, pp 1–83. https://doi.org/10.1016/B978-0-12-405890-3.00001-4

Kumar A, Sharma M (2020) Sustainable development of riverine islands: a case study from Eastern India. Sustain For 12(15):6142. https://doi.org/10.3390/su12156142

Kumari A (2022) Chapter 13 – riverine biodiversity and importance: Potential threat and conservational challenges. In: Madhav S, Kanhaiya S, Srivastav A, Singh V, Singh P (eds) Ecological significance of river ecosystems. Elsevier, pp 235–264. https://doi.org/10.1016/B978-0-323-85045-2.00009-1

Mishra SK (2019) Participatory resource management in riverine communities. Environ Manag 63(4):415–428. https://doi.org/10.1007/s00267-019-01140-3

National Bureau of Soil Survey and Land Use Planning (2015) Soil health and fertility. Indian Council of Agricultural Research. Available at: https://nbsslup.in/soilhealthfertility

Rath S (2018) Traditional ecological knowledge in natural resource management. Environ Conserv 45(4):350–358. https://doi.org/10.1017/S0376892918000115

Rout N (2020) Eco-tourism potential in Odisha: opportunities and challenges. Tour Manag Perspect 33:100616. https://doi.org/10.1016/j.tmp.2019.100616

Singh R (2018) Riverine ecosystems in India: challenges and conservation strategies. Environ Manag 62(1):1–15. https://doi.org/10.1007/s00267-018-1012-x

Tiwari AK, Singh R, Kumar S, Singh GS (2023) Ecosystem services in the riverine landscapes. In: Rai PK (ed) Advances in water resource planning and sustainability. Advances in geographical and environmental sciences. Springer, Singapore. https://doi.org/10.1007/978-981-99-3660-1_16

Soil Genesis Classification and Land Use of Some Soils in the Upper Mahanadi Basin

4

Antaryami Mishra, Ajay Kumar Mishra, and Prava Kiran Dash

Abstract

The Mahanadi River Basin covers approximately 75% of the geographical area of the state of Odisha and spans 23 districts. Its ecological balance depends on the preservation of the soil and forest resources of these districts. The entire area consists of hilly tracks and undulating topography with distinct toposequences. Hence, a detailed study of the soil morphology, genesis and classification of riverside areas is used as the basis for the preservation of its ecosystem.

In this present investigation, selected pedons representing different land types in a toposequence in the upper Mahanadi basin were studied in detail, with the land types being hill slope land, upland land, medium land, lowland land and stream terrace land, which represent the ideal topography of the basin. The leaching of basic cations from the A to B horizons in the process of *eluviation* and *illuviation* is well evidenced. Out of the 12 orders in *soil taxonomy*, *Alfisols*, *Inceptisols*, *Entisols* and *Vertisols* are present here.

A. Mishra (✉)
Department of Soil Science and Agricultural Chemistry, Institute of Agricultural Sciences (IAS), Siksha 'O' Anusandhan Deemed to be University, Bhubaneswar, Odisha, India
e-mail: a.k.mishra@irri.org

A. K. Mishra
International Rice Research Institute, South Asia Regional Centre, Varanasi, Uttar Pradesh, India

P. K. Dash
Department of Soil Science and Agricultural Chemistry, Institute of Agricultural Sciences (IAS), Siksha 'O' Anusandhan Deemed to be University, Bhubaneswar, Odisha, India

Regional Research and Technology Transfer Station, Dhenkanal, Odisha, India

J. P. Bhatt, A. K. Mishra (eds.), *Mahanadi River*,
https://doi.org/10.1007/978-981-96-1048-8_4

The hill slopes and uplands are suitable for forest species, timbers and plantation crops such as sisal for soil conservation, cashew, mango, guava, custard apple, litchi, orange, drumstick, plumeria, pineapple, etc. The medium lands are suitable for vegetables, pulses and oilseeds. The lowlands are most suitable for growing rice.

Keywords

Mahanadi Basin · Odisha · Land types · Soil classification · Land use

Mahanadi, one of the most important rivers in India, originates from the Chhattisgarh state and then enters the Odisha state in Jharsuguda district. Then, it flows through the Sambalpur, Boudh, Dhenkanal, Cuttack and Jagatsinghpur districts and falls into the Bay of Bengal. It is the lifeline of the Odisha state. The river basin of Mahanadi, from which water drains to this important river through various tributaries, is spread in 23 districts of the state, occupying approximately 75% of the total geographical area of the state. The districts directly influenced by the Mahanadi River are Jharsuguda, Sambalpur, Bargarh, Subarnapur, Boudh, Kandhamal, Angul, Cuttack, Jagatsinghpur, Kendrapara, Khordha and Puri. The districts of Jharsuguda, Sambalpur, Bargarh, Subarnapur and Boudh constitute the upper Mahanadi basin, which consists mostly of hilly tracks and undulating topography with distinct toposequences. Hence, an in-depth study of the soil morphology, genesis, characterization and classification of the soils of a riverbank area forms the real basis of preservation of its ecosystem.

The present study was undertaken on the left bank of the river in ideal toposequences, which are representative of the basin. This area falls under the Hirakud Command Area, which is the total irrigation area by canals originating from the Hirakud Dam Project, which is constructed across the Mahanadi River in the district of Sambalpur.

The Hirakud Dam Project on the Mahanadi River is one of the major irrigation projects in India. It has a gross command area of approximately 7 lakh acres. The climate of the command area is subtropical, monsoonal, hot and moist subhumid, with pronounced alternating wet and dry periods throughout the year. The mean annual rainfall is 1456 mm, and the mean annual air temperature is 26.7 °C. The soil temperature is *hyperthermic*.

The area belongs to the Gondwana Shield low tableland and humid landform. The area has five prominent landforms, namely hills, ridges, valleys, levees and stream terraces, and a good assemblage of sequential landforms in toposequences. The basic drainage patterns of the area are *diverse*.

Biotite genesis is the dominant rock type in the study area. The rock systems are rich in ferromagnesian minerals. The lithology of the area also comprises *alluvium*, which differs considerably in its age and extent.

Starting from the flat upland at the highest topography to the low levee top land on the riverbank of Mahanadi, seven pedons (soil profiles) were identified in a toposequence for the study. These profiles are assigned the names of the seven soil series, which are the names of nearby villages where they occur on the left bank of the Mahanadi River.

These soil series in a toposequence starting from the highest elevation to the riverbank are as follows: (1) Putiapali (flat upland), (2) Radhiapali (upper valley side land), (3) Batemura (lower valley side land), (4) Chakuli (upper flat valley bottomland), (5) Dhama (high levee slope land), (6) Baunsara (high levee top land) and (7) Sankhala (low levee slope land). These pedons are denoted Pedons 1, 2, 3, 4, 5, 6 and 7, respectively.

All the soils under these toposequences are predominantly of *fluvial* origin (Mishra 1981; Mishra et al. 2019). Except for upper flat valley bottomland (Pedon 4 represented by the Chakuli series), the soils of the remaining land types were initially subjected to the geogenic process of *alluviation* and subsequently to considerable erosion of different intensities, depending on their physiographic position. The depth of the solum at the ridge—top—has a flat upland at the minimum, which increases in soils down the slope and becomes more than 2 m deep in the upper flat valley—bottomland. The slope from the upland to the valley–bottomland is very gentle. All the land types in the levee have soils thicker than 2 meters. While the great depth of soils in the levee could be attributed to the great depth of deposition of materials through the process of *alluviation*, the lower depth of soils in the upland and upper parts of the upper valley–side land is predominantly due to *erosion* (Sahu and Mishra 1994; Mishra et al. 2009).

The soils have been cultivated for more than a thousand years and have been subjected to considerable man-made erosion, resulting in the removal of materials from the upper parts of the slope and their deposition on the lower part (Jenny 1941).

The soil colour becomes yellower down the slope both at the surface and in a major part of the subsoil, which could be attributed to the increasing moisture regime and impeded drainage (Russel 1961). The development of grey colour could also be due to the combined effects of high clay content and high calcium and moisture status (Gawande 1961; Lokya et al. 2021).

The texture of the soils both at the surface and below gradually becomes heavier down the slope. In particular, the climatic conditions prevailing in areas with high rainfall cause considerable erosion in the upland and upper slopes, and the runoff carrying the soil materials is deposited in the lower part of the slope and in the valley bottom. The very gravelly and gravelly conditions in the subsoils of Pedon 1 and Pedon 2, respectively, are due to the presence of rounded pebbles and cobbles in addition to ferruginous nodules and the loss of finer particles through runoff (Dash et al. 2022a, b). Along the slope, the soil structure also becomes better developed in size, grade and form, which could be attributed chiefly to the higher clay content of the soils of the subland types at the lower slopes. However, although Pedons 6 and 7 have lower clay contents than Pedons 1 and 2 do, they have better structures, which could be due to the lower gravel content of the two former pedons. In Pedon 4, the peculiar wedge-shaped structural aggregates are due to the process of *argillipedoturbation* (Sahoo 2020).

The abundant medium ferruginous concretions in Pedon 1, which have considerably contributed to the coarse fragments in addition to the rounded pebbles and cobbles to make the soil skeletal, have been formed under alternating wet and dry conditions and favourable flat topographic conditions, with the underlying bedrock considerably restricting the downward movement of water in the wet season. The mottles observed in all the soils are mostly due to climatic effects that create alternating reducing and oxidizing conditions (Biswas and Gawande 1962; Buol et al. 1980).

The occurrence of calcium concretions in Pedons 3, 4 and 5 could be attributed to the higher content of clay rich with calcium, inherited from materials from which they have formed and supplemented with that received from the slopewash from the land types and higher physiographic positions. The percolation of water is greatly restricted by the high clay content in all the subhorizons of Pedon 4 and the lower parts of Pedons 3 and 5. In addition, Pedon 3 has underlying bedrock that restricts the downwards movement of water along with bases, including the most predominant ion, calcium. The absence of calcium concretions in Pedon 6 could be due to the initially low content of calcium in the alluvial deposit from which the solum has developed and to the highly leached conditions of the soil for a longer period.

The presence of broken, moderately thick clay skins in all the Pedons except Pedons 1 and 4 indicates favourable conditions for development. The poor development of clay skin in Pedon 1 is due to the poor development of the structure itself, on which the clay skins could develop. The absence of clay skin in Pedon 4 is due to the process of *argillipedoturbation* (Mohanty 2012).

As shown in the table, the percentage of clay, cation exchange capacity (CEC), base saturation percentage (BSP) and pH of all the pedons tended to increase along the slope, except for a minor variation in the pH of the surface soil. The physical and chemical characteristics are thus in toposequences (Mishra et al. 2019) and could chiefly be attributed to the transportation of finer soil particles along with bases down the slope through runoff and their deposition in lower physiographic positions. Kalpage (1974) reported that the loss of clay in surface horizons downslopes was due to both lateral and vertical eluviation.

All the soils of this toposequence in the upper Mahanadi basin, which is representative of the entire basin because of its typical topographic position, are placed under the soil order *Alfisols*, which formed particularly owing to favourable leaching conditions to develop an *argillic* horizon with moderate to high base status. The soil represented by Pedon 4, which is placed under the order *Vertisols,* separates out at the order level.

The high content of clay inherited from the sediments, deposited while plugging sloughs and received through calcium- and magnesium-rich slopewash from the upland and valley-side land, has been quite favourable under alkaline conditions for the formation of *montmorillonite*. Thus, the high contents of clay and montmorillonite, as indicated by high CECs, have resulted in the formation of *Vertisols*, which are known to exhibit good characteristics.

Among the *Alfisols*, Pedon 3 belongs to *Aqualf,* which is separated at the suborder level from the other five Pedons, which are *Ustalfs*. The greater moisture regime resulting in the formation of *Aqualfs* could be due to considerable moisture received from the upland and upper ends of the valley-side land and their restricted percolation. The soil represented by Pedon 6 is a Paleustalf and separates at the great group level from Pedons 1, 2, 5 and 7, which are *Haplustalfs*. The favourable conditions of the very old and stable levee top, particularly the wide flat topography and very deep water table, have caused greater translocation of clay particles from the A horizon to the B horizon (Jenny 1941), as evident from the high ratio of clay in the B/A horizon. The different chemical properties mentioned in the table, in addition to the clay distribution and the presence of many coarse yellowish red mottles in this pedon, justified its grouping under *Paleustalfs*. (All the soil colour notations given in the table are per the '*Munsell soil colour chart*'.)

Out of the four pedons of *Haplustalfs*, Pedon 1 separates at the family level due to loamy-skeletal particle size, which is chiefly due to fluvial deposition of rounded pebbles when the area formed a part of the Mahanadi River and subsequent *laterization* and erosion as the river gradually reached lower positions.

The soils represented by Pedons 2, 5 and 7 differ at the series level and are named Radhiapali, Dhama and Sankhala, respectively. Pedon 2, which represents the Radhiapali series, has developed under conditions nearly similar to those of Pedon 1 but is located in the upper part of the upper valley–lateral land; because of higher slopes, *laterization* has not been intense. The coarse fragment content, though more than 20%, is less than 35% in the B horizon. The subsoil is gravelly but not skeletal in the family control section for particle size. However, having more than 20% of the rounded pebbles in the lower part of the B horizon over weathered bedrock of biotite gneiss, the Radhiapali soils represented by Pedon 2 differ from those of the Dhama and Sankhala soils, which are represented by Pedons 5 and 7, respectively.

In terms of clay distributions at different horizons, within the series control section, Pedon 7 is loamy in all parts, with chroma of more than 2, whereas Pedon 5, in its lower part within the series control section, is distinctly clayey and has a chroma of 2 due to wetness..

Among all the pedons in the toposequence, Pedon 6 has the lowest CEC and BSP, which could be attributed to the high leaching conditions and the low pH resulting from base depletion. Pedon 4 not only has the maximum clay content, CEC and BSP but also has fairly homogenous properties within the soil solum because of churning, which has caused sufficient vertical mixing of materials to a considerable depth.

Thus, the most contrasting and interesting soils of this toposequence from the point of view of pedogenesis are Baunsara (Pedon 6) and Chakuli (Pedon 4), which are located in the high levee top land and upper flat valley–bottomland, respectively. As may be observed from their morphological, physical and chemical characteristics, while the former has been subjected to high leaching and *argillation,* exhibiting distinct horizonation, the other soil has been significantly homogenized in the solum through the process of *argillipedoturbation* (Mishra et al. 2019).

Thus, the differences in soil genesis and soil taxonomy, even at the micro level, in the upper Mahanadi basin indicate that different soil management systems should be used to maintain soil health, check for soil erosion and preserve the ecosystem of the river basin for a brighter and environmentally friendly future.

With respect to soil management systems, hill slope lands and uplands are suitable for growing forest species; timbers; plantation crops suitable for soil conservation, such as sisal (*Agave sisalana*) (Mishra and Sahu 1993); and fruit trees, such as cashew, mango, guava, custard apple, litchi, orange, drumstick and pineapple.

Medium lands are suitable for growing all types of vegetables, pulses and oilseeds. The lowlands are most suitable for growing rice. Utilization of different land types on the basis of their soil characteristics and crop suitability (Mishra 2007) for forestry, silvopastoral systems, orchards and field crops, starting from the hill slope to the riverbed through the toposequence from the highest elevation to the lowest elevation in the hills and undulating upper Mahanadi basin, will help in preserving the ecosystem, conserving soils in the riverbed and banks of different tributaries of the Mahanadi River, which will ultimately help in assessing siltation and enhancing the productivity of crops (Table 4.1).

Table 4.1 Important morphological, physical and chemical characteristics of the soils of a toposequence in the upper Mahanadi basin and differences in their taxonomy

Characteristics	Pedon 1	Pedon 2	Pedon 3	Pedon 4	Pedon 5	Pedon 6	Pedon 7
Thickness of the solum	107 cm	119 cm	175 cm	200 cm	> 200 cm	> 200 cm	> 200 cm
Taxonomic classification							
Order	Alfisol	Alfisol	Alfisol	Vertisol	Alfisol	Alfisol	Alfisol
Suborder	Ustalf	Ustalf	Aqualf	Ustert	Ustalf	Ustalf	Ustalf
Great group	Haplustalf	Haplustalf	Ochraqualf	Chromustert	Haplustalf	Paleustalf	Haplustalf
Subgroup	Udic Haplustalf	Udic Haplustalf	Aeric Ochraqualf	Udorthentic Chromustert	Udic Haplustalf	Udic Paleustalf	Udic Haplustalf
Family							
Particle size	Loamy-skeletal	Fine-loamy	Fine	Fine	Fine-loamy	Fine-loamy	Fine-loamy
Mineralogy	Mixed	Mixed	Montmorillonitic	Montmorillonitic	Mixed	Mixed	Mixed
Series	Putiapali	Radhiapali	Batemura	Chakuli	Dhama	Baunsara	Sankhala
Morphological characteristics							
Colour (as per the Munsell colour chart)							
Surface soil	10YR5/4	2.5YR5/4	2.5Y4/2	2.5Y4/2	2.5Y5/2	10YR6/2	10YR6/4
Subsoil	10YR5/4	2.5Y4/4	5Y5/2	5Y4/2	5Y5/2	10YR5/2	5Y5/3
Texture							
Surface soil	Sandy loam	Sandy loam	Sandy clay loam	Clay	Loam	Sandy loam	Sandy loam

(continued)

Table 4.1 (continued)

Characteristics	Pedon 1	Pedon 2	Pedon 3	Pedon 4	Pedon 5	Pedon 6	Pedon 7
Subsoil	Very gravelly sandy clay loam	Gravelly sandy clay loam	Sandy clay	Clay	Sandy clay	Sandy clay loam	Sandy clay loam
Structure							
Surface soil	Fine weak subangular blocky	Fine weak subangular blocky	Medium-strong subangular blocky	Coarse strong subangular blocky	Fine weak subangular blocky	Fine weak subangular blocky	Fine weak subangular blocky
Subsoil	Fine weak subangular blocky	Medium-moderate subangular blocky	Coarse strong angular blocky	Wedged-shaped aggregate breaking into coarse strong angular blocky	Coarse strong angular blocky	Coarse moderate angular blocky	Coarse moderate angular blocky
Mottling							
Surface soil	Prominent many fine 7.5YR6/8	Prominent many fine 7.5YR5/6	Prominent many fine 7.5YR5/6	Prominent many fine 7.5YR7/6	Prominent common medium 7.5YR5/6	Prominent common medium 7.5YR5/6	Prominent many fine 7.5YR5/6
Subsoil	Faint a few fines 7.5YR4/4	Faint a few fine 5YR5/6	Distinct few fine 2.5Y4/4	Faint few fine 7.5YR5/6	Faint few fine 7.5YR5/6	Prominent many courses 5YR4/6	Distinct common medium 7.5YR5/6
Concretion	Abundant medium iron concretions	Plentiful medium iron concretions	Plentiful medium iron and calcium concretions	Plentiful medium iron and few calcium concretions	Plentiful medium iron and coarse calcium concretions	Plentiful medium iron concretions	Plentiful medium iron and few medium calcium concretions
Clay skin	Patchy thin	Broken moderately thick	Broken moderately thick	Nil	Broken moderately thick	Broken moderately thick	Broken moderately thick

Physical and chemical characteristics							
Clay (percent)							
Surface soil	14.5	17.1	24.7	45.6	11.0	5.1	9.6
Subsoil	33.2	34.1	38.9	50.1	37.0	23.9	28.9
CEC (cmol(p^+) kg^{-1}							
Surface soil	12.46	13.19	16.89	33.87	9.89	3.96	5.80
Subsoil	21.17	24.19	34.12	45.87	25.72	11.72	15.49
Base saturation (percent)							
Surface soil	71.91	84.84	87.83	89.67	77.08	49.49	74.14
Subsoil	80.59	86.96	92.84	95.65	91.14	71.05	85.80
pH							
Surface soil	6.2	6.4	6.6	6.2	6.7	5.8	6.1
Subsoil	6.4	6.8	7.2	7.5	7.4	6.1	6.5

References

Biswas TD, Gawande SP (1962) Studies in genesis of catenary soils on sedimentary formation in Chhattisgarh basin of Madhya Pradesh. I-morphology and mechanical composition. J Indian Soc Soil Sci 10:223–234

Buol SW, Hole FD, McCracken BJ (1980) Soil genesis and classification. Oxford and IBH publishing Co., New Delhi

Dash PK, Mishra A, Patro H, Nayak RK, Jena B, Satpathy M (2022a) Soil characterization along a toposequence located in north-central plateau Agroclimatic zone of Odisha. Ecol Environ Conserv 28:471–476

Dash PK, Mishra A, Patro H, Sahoo BC, Panda RK, Saren S (2022b) Morphological characterization of soil profiles in a toposequences located in the north eastern coastal plains agroclimatic zone of Odisha. Environ Ecol 40(3C):1762–1772

Gawande SP (1961) Pedological studies of development of associated soils of toposequence. Ph.D. Thesis. Division of Soil Science and Agricultural Chemistry, IARI, New Delhi

Jenny H (1941) Factors of soil formation, a system of quantitative pedology. McGraw-Hill. Book Co. Inc, New York

Kalpage FSCP (1974) Tropical soils: classification, fertility and management. The MaCmillan Co. of India Ltd, Calcutta

Lokya T, Mishra A, Saren S (2021) Characterization and taxonomic classification of soils of hill slope land and stream terrace land located at North–Eastern Ghat Agroclimatic zone of Odisha, India. Int J Plant Soil Sci 33(17):203–209

Mishra DP (1981) Morphological studies and classification of soils of Hirakud command area. Ph.D. Thesis. Dept. of Soil Science and Agricultural Chemistry. Orissa University of Agriculture and Technology, Bhubaneswar

Mishra A (2007) Land suitability classification for different crops. Orissa Rev LXIV(4):65–68

Mishra A, Sahu GC (1993) Sisal-the strongest vegetable fibre crop. Orissa Rev XLIX(10):55–56

Mishra A, Sahu GC, Nanda SK, Muralidharudu Y (2009) Vertical distribution of available potassium in some soils of west central table land catena in Orissa, India. In: Proceedings IPI-OUAT-IPNI international symposium, vol 2, Bhubaneswar, pp 27–30

Mishra A, Saren S, Dash PK (2019) Soil formation in Toposequences – a review. Lambut Academic Publishing

Mohanty R (2012) Preparation of GPS based soil fertility maps and identification of soil fertility maps and identification of soil related crop production constraints for Angul district, Odisha. M.Sc. (Ag) Thesis. Department of Soil Science and Agricultural Chemistry, OUAT, Bhubaneswar

Russel EW (1961) Soil conditions and plant growth, 9th edn. Longmans, Green and Co. Ltd, London

Sahoo CS (2020) Preparation of GPS based soil fertility maps and identification of soil related crop production constraints of RRTTS and KVK farm, Bhawanipatna. M.Sc. (Ag) Thesis. Department of Soil Science and Agricultural Chemistry, OUAT, Bhubaneswar

Sahu GC, Mishra A (1994) Morphology, characteristics and classification of soils under sisal (*Agave sisalana*) cultivation. J Indian Soc Soil Sci 42(1):111–114

Why Is the Mahanadi River Not a Conducive Habitat for Large Reptiles, Especially Crocodilia?

5

Utpal Mohanta, Anannya Das, Abheepsha Priyadarshini Pradhan, and Satyajeet Arya

Abstract

The Mahanadi River ecosystem is home to three crocodilian species in India: the gharial, mugger, and saltwater crocodile. However, their populations have significantly decreased in recent years, prompting urgent conservation efforts. Thus, present study aims to identify the drivers that threaten the populations of crocodilian fauna in the Mahanadi River and to examine the conservation initiatives of governmental agencies. This contribution also provides a comprehensive overview of the natural history of crocodilians, including their distribution, habitat requirements, feeding habits, reproductive patterns, behaviors, and the threats they face. The discussion extends to future conservation strategies, emphasizes the importance of habitat restoration, strengthens protection of designated areas, addresses the community engagement, and highlights sustainable river management practices. This chapter concludes that the long-term viability of these reptile species in the Mahanadi ecosystem hinges on sustained collaboration among governmental agencies, conservation organizations, local communities, and researchers.

Keywords

Mahanadi River · *Gavialis gangeticus* · *Crocodylus palustris* · *Crocodylus porosus* · Threats · Conservation

U. Mohanta · A. Das · A. P. Pradhan · S. Arya (✉)
Environmental Science, Faculty of Science, Sri Sri University, Cuttack, Odisha, India

J. P. Bhatt, A. K. Mishra (eds.), *Mahanadi River*,
https://doi.org/10.1007/978-981-96-1048-8_5

5.1 Introduction

The crocodilian fauna of India comprises three species: the Indian gharial (*Gavialis gangeticus*), the mugger or marsh crocodile (*Crocodylus palustris*), and the saltwater crocodile (*Crocodylus porosus*) (Saikia 2013). These three species are commonly referred to as *makara* in the Indian literature (Ganesan 2007). Gharial is the only surviving member of the family Gavialidae and has a highly restricted distribution in the rivers of India and Myanmar (Saikia 2013). On the other hand, the salt crocodile is more widely distributed than the two other species under discussion. All three species are listed under Schedule I of the Wildlife Protection Act of 1972, granting them full legal protection at all times (Basu et al. 2023). These species face varying levels of threat, with the gharial classified as "critically endangered" and the mugger as "vulnerable" on the IUCN Red List (Webb et al. 2019).

Before the 1970s, all three crocodilian species experienced drastic declines in population due to several factors. The gharial, once abundant in northern India's rivers, faced near extinction by the middle-twentieth century due to habitat destruction, poaching, and the detrimental effects of fishing practices (Panda et al. 2023). The saltwater crocodile was wiped out in several Indian states by 1974, with only small populations persisting in selected areas, including the Brahmani–Baitarani deltaic region of Odisha (Lloyd-Jones et al. 2023). Similarly, the mugger, despite its once widespread distribution, has faced substantial declines across its range, primarily due to habitat loss, human–wildlife conflict, and entanglement in fishing gear (Sharma et al. 2021).

The Mahanadi River ecosystem in Odisha, India, provides unique habitats for all three Indian species of crocodilians. However, population estimates are unavailable, and the status of all three species in Mahanadi is highly precarious. Unlike fish species, crocodiles are not overexploited for livelihood purposes. Nevertheless, factors such as river regulation, habitat destruction, sand mining, riverbank encroachment, poaching, and other human activities have contributed to the declining populations of these large reptiles (Chen et al. 2023; Lamichhane et al. 2022; Nawab 2013; Chaudhari et al. 2022; Panda et al. 2023). These detrimental factors are prevalent in the Mahanadi River and are likely responsible for the diminishing crocodilian populations in this river system.

In this contribution, we provide an overview of some aspects of the natural history of all three species, to enhance our understanding of the threats to and conservation of the crocodilian fauna in the Mahanadi River system.

5.2 Geographic Distribution

Among the three crocodilian species, saltwater crocodiles (*Crocodylus porosus*) are the most widely distributed. Its range spans South Asia and Southeast Asia. In India, it can be found along the eastern coast, in the Andaman and Nicobar Islands, and in the Sundarbans mangrove habitats of West Bengal (Kar 2013; Singh and Kar 2006; Webb et al. 2019).

The mugger crocodile (*Crocodylus palustris*) is indigenous to the freshwater regions of South Asia, with a range covering countries such as Iran, India, Nepal, Pakistan, and Sri Lanka (Daniel 2002). In India, this species inhabits various protected areas, including Kuno Wildlife Sanctuary (Madhya Pradesh), Katarniaghat Wildlife Sanctuary, Simlipal Tiger Reserve, and Satkosia Gorge Wildlife Sanctuary (Odisha) (Dutta et al. 2009; Das et al. 2021). Outside the protected areas, its presence has been recorded in the Chambal River (Madhya Pradesh and Rajasthan freshwater bodies in the states of Karnataka, Tamil Nadu, and Gujarat; the Ghodahada Reservoir in the Ganjam district and the Sabari River in the Koraput district of Odisha (Sagar and Singh 1993; Debata et al. 2018).

The distribution of gharial (*Gavialis gangeticus*), which inhabits the freshwater systems of India and Myanmar, is much more restricted. Historical reports suggest that the range of gharial once covered the Indus River in Pakistan to the Irrawaddy River in Myanmar. Currently, it is confined to a few natural water bodies, such as the Chambal River, Girwa River, Son River, Brahmaputra River, and Rapti and Narayani rivers (Saikia 2013).

Recent newspaper reports indicate that crocodilian species, particularly the Indian gharial and mugger, have been detected in various parts of the Mahanadi River. In the past, researchers documented the exclusive presence of the gharial in the Indian subcontinent, including the Mahanadi River system (Whitaker 1987). However, it is evident that the water conditions of the Mahanadi River may not support self-sustaining populations of crocodilian species.

5.3 Habitat

Crocodiles primarily inhabit tropical environments within their geographic ranges in Asia, Africa, the Americas, and Australia. Mugger crocodiles typically prefer rivers, lakes, and marshes; however, they have also been reported in reservoirs, irrigation canals, man-made ponds, hill streams, coastal saltwater lagoons, and estuaries (Whitaker 1987; Whitaker and Whitaker 1989). The saltwater crocodile primarily inhabits coastal saltwater regions but also utilizes freshwater rivers and marshes as feeding grounds. This species exhibits considerable tolerance to salinity. They are typically observed in brackish water environments surrounding coastal zones and rivers. However, they are also frequently encountered in swamps, billabongs, and other freshwater bodies (Newman 2023). The gharials usually live in fast-moving rivers with significantly deep pools, elevated sand banks, and abundant fish populations (Whitaker and Basu 1981). They live at depths of confluence in river junctions and groups, known as "kunds" or bare and steep ravines in mountainous terrain. It is believed that they disperse widely in monsoon floodwaters and return to enduring water holes after the rainy season.

The lower reaches of the Mahanadi River seem conducive for all three species of crocodiles, as it has a rich freshwater drainage system, lakes, estuaries, and saltwater lagoons; however, the current situation does not support these findings. It is mentioned by researchers that the Mahanadi River does not provide a suitable environment for large reptiles such as crocodilians because it is subjected to numerous human-induced pressures and ecological stresses. The swift pace of industrial growth, the use of coal for electricity production, and the excessive diversion of water to industrial activities in regions such as Odisha and Chhattisgarh have significantly affected the current and overall conditions of the river basin and increased its susceptibility to climate change. Therefore, the required habitat is destroyed and becomes nonconducive to crocodilians.

5.4 Food and Feeding Habits

All three species of crocodiles under discussion are carnivorous. The diet of adults includes fish, turtles, birds, and small mammals, whereas hatchlings and smaller individuals feed on insects, snails, and crabs. However, researchers have also observed slight variations in the feeding behaviors of these species. Murthy and Pillai (1990) reported that gharials cannot cut or chew objects with their teeth, which are specifically adapted for holding prey. Nonetheless, they possess the ability to swallow large objects because of their wider throat. If the prey is excessively large and unable to be swallowed in its entirety, the species employs a macabre technique involving the dismemberment of the victim limb by the limb, accomplished through the rapid rotation of its entire body. The muggers sometimes forage for food outside water bodies. There are instances where large adults were reported to prey on deer and buffalos (Whitaker and Whitaker 1989).

Saltwater crocodiles are generally nocturnal and can attack any animal they manage to catch, including humans and other estuarine crocodiles. They move from under the water surface to attack an unaware target, dragging it into the water to consume it (Rafferty 2024).

5.5 Breeding Habit

In mugger crocodiles, brooders reach sexual maturity when they reach a length of approximately 1.8–2.0 m. Females lay 25–30 eggs during the early dry season (Whitaker and Whitaker 1989). Nesting occurs in May, with hatching typically occurring in July. In hatcheries, mugger crocodiles have been reported to produce two clutches annually. The incubation period usually lasts for a duration of 55–75 days (Whitaker 1987).

Saltwater crocodiles reach sexual maturity at approximately 12 years for females and 16 years for males. Females build mound nests from vegetation between November and May, laying an average of 50 eggs. Temperature is known to influence the sex ratio of saltwater crocodiles. For example, at 28 °C to 30 °C, all eggs hatch as females; at 31 °C, the sex ratio is equal; and at 32 °C, all eggs hatch as males. The proportion of females ranged from 50% to 100% at temperatures between 33 °C and 34 °C. The incubation period for saltwater crocodiles is approximately 75 days (Nair 2019).

For gharials, females and males reach sexual maturity at sizes of approximately 2.5 m and 3.0 m, respectively, typically between 13 and 14 years of age. Mating takes place in December and January, with nesting occurring from late March to early April, generally with a variation of no more than 10 days. Clutch sizes range from 10 to 60 eggs. The optimal temperature for breeding ranges from 32 to 34 °C, and the incubation period lasts between 72 and 92 days, with an average of 84.5 days. Sand is the preferred spawning ground of gharials, although they also lay eggs in muddy soil and alluvial deposits (Rao 1988).

5.6 Behavior

To regulate body temperature, mugger crocodiles rely on both water and land. They practice heliothermy by basking in the sun and thigmothermy by absorbing heat from warm surfaces (Patil et al. 2012). In hot weather, they dig burrows to cope with high temperatures (Da Silva and Lenin 2010). Muggers hunt by remaining motionless in the water, waiting for prey to come within a striking distance, and then launching a sudden attack. Additionally, they hunt on land, sometimes running at speeds of up to 13–14.5 km (8–9 miles) per hour to catch their prey (Rafferty 2024).

Saltwater crocodiles are fast swimmers and voracious feeders. They can swim at speeds of 12–15 km per hour and attack anything that fits in their mouths. They are agile both on land and in water, with an excellent sense of hearing that aids in locating prey. Saltwater crocodiles can thrive in freshwater environments, with some individuals found as far as 235 kilometers from the ocean (Newman 2023). They possess unique communication abilities, where each "bark" conveys different meanings. They mark their territory by secreting scents, and while they can bask on both water and land surfaces, they spend more time in water than other species do (Kumar et al. 2012).

Gharials, with their long snouts and rows of uniformly sharp teeth, are highly adapted to capture fish, which is their primary diet. Their long snouts make them the most efficient fish hunters among the crocodilian species in Indian waters. Gharials are predominantly located on rocky surfaces, particularly during winter. They tend to use the same basking locations, increasing their vulnerability (Singh and Bustard 1977a, b). Compared with mugger crocodiles, they also bask for longer periods.

5.7 Threats

Crocodilian fauna are considered top predators in aquatic ecosystems, especially rivers and lakes. They play crucial roles in maintaining the structural and functional integrity of freshwater ecosystems. These large and ancient reptiles face survival challenges worldwide because of human activities and certain intrinsic factors. The populations of crocodiles under discussion are highly vulnerable in the Mahanadi River system. Rapid industrialization, coal-fired power generation, and excessive water allocation to industries in Odisha and Chhattisgarh have significantly affected river flow, basin health, and susceptibility to climate change. Ratha (2019) reported that anthropogenic stresses and ecological disasters have made the Mahanadi River an unsuitable habitat for large reptiles such as crocodiles. Inadequate environmental flow in various stretches of the Mahanadi River, caused by river regulation, has led to adverse impacts on the integrity of the ecosystem. This habitat degradation has not only negatively affected large species such as crocodiles but also decreased their food sources, such as fish and crabs, due to excessive fishing and insufficient water flow (Vashistha et al. 2021). The construction of the Hirakud Dam on the Mahanadi River in 1957 marked the beginning of extensive dam construction in the region. This has resulted in a reduced sediment supply and increased coastal erosion, which has likely altered crocodile habitats (Mukhopadhyay et al. 2018). Since the dam's construction, the sightings of gharials and muggers have significantly declined (Somanna et al. 2016).

In addition to flow modification, fishermen's use of nylon set netting, increased demand for eggs, nest predation, poaching, hunting for meat, and hide trading are the factors responsible for the declining populations of crocodiles, especially gharials (Marshall et al. 2020; Chavan and Borkar 2023). Da Silva and Lenin (2010) identified accidental drowning in fishing nets as a significant threat to muggers.

Increased urbanization and industrialization have led to higher levels of untreated sewage and industrial effluent in the Mahanadi River system (Sundaray et al. 2006). This has resulted in organic contamination, low dissolved oxygen levels, and increasingly acidic conditions (Panda et al. 2006; Behera et al. 2014). The increasing presence of plastics in sediments and water bodies threatens fish, marine animal, and bird populations (Hazra et al. 2020). Large-scale dumping of municipal waste and industrial waste has a detrimental impact on riverine ecosystems. Muggers have lost basking sites due to concentrated waste, potentially encouraging them to relocate or leave the territory and increasing the likelihood of human–mugger interactions (Choudhary et al. 2018).

Sand mining in the Mahanadi River is also one of the factors contributing to the decline in crocodile populations. This activity not only destroys their breeding grounds but also exposes incubated eggs (Ratha 2019). In addition to human-induced activities, certain intrinsic characteristics of these large reptiles, such as their long incubation period, make their populations vulnerable. Eggs are not only destroyed by sand mining but also eaten by terrestrial predators such as jackals (Whitaker and Basu 1981).

5.8 Conservation Measures

The conservation of large reptiles requires a comprehensive understanding of their ecology and natural history (Williamson et al. 2017). Globally, extensive research and conservation efforts have been conducted, including habitat restoration and antipoaching initiatives, particularly in Southern Africa, America, and Asia. Similar exercises were practiced in various states of India, such as Madhya Pradesh, West Bengal, and Gujrat. The National Chambal Sanctuary in Madhya Pradesh has implemented several measures, particularly those focused on gharials and muggers, by establishing captive breeding programs, providing habitat protection from sand mining activities, and engaging local communities in conservation efforts through awareness programs and education (Sharma and Singh 2015). Similar efforts were made in the Bhagabatpur reserve forest in West Bengal and the Kachchh desert sanctuary in Gujrat for muggers (Adhya 2012; Desai et al. 2024).

In Odisha, conservation measures for mugger crocodiles have resulted in positive developments, particularly in the Simlipal Tiger Reserve and the Satkosia Gorge Wildlife Sanctuary. The successful implementation of programs such as captive breeding and rehabilitation programs has resulted in a multifold increase in the number of released muggers (Sagar and Singh 1993).

A similar practice was created by the Bhitarkanika Crocodile Conservation Project, located in Bhitarkanika National Park in Odisha, India, which was a significant initiative aimed at the conservation of estuarine or saltwater crocodiles (Khan et al. 2020). The project to conserve crocodiles in Bhitarkanika started in 1975 because there was a significant decrease in the crocodile population caused by habitat destruction, poaching, and human interference. A hatchery was established at Dangmal, which is located within Bhitarkanika National Park, and the project adopted a "rear and release" approach. Crocodile eggs were collected from the wild, and the young were reared away from predators. Furthermore, upon reaching a suitable size, they were released into protected areas. Conservation efforts have led to a significant increase in the crocodile population of Bhitarkanika. From just a few individuals in the 1970s, the population has grown to over 1600 individuals, which also contributed to the overall health of the mangrove ecosystem (Gopi and Pandav 2009). The "rear and release" program initiated in 1975 has been particularly successful, with over 2950 captive-reared juveniles released between 1977 and 2015.

To strengthen the conservation efforts for the gharials, a Gharial Research and Conservation Unit (GRACU) was established in Tikarpada, Odisha, which plays a significant role in the early conservation efforts of the species. Over the past few decades, these programs have yielded mixed results. While some populations have shown signs of recovery, others continue to face challenges. A recent study by Rath et al. (2024) on the "Reintroduced Gharials" in the Mahanadi River between 2019 and 2021 highlighted a significant increase in the population within the Satkosia Gorge Wildlife Sanctuary. However, these populations still face threats from human activities such as overfishing and entanglement in fishing nets.

Nandankanan Zoological Park's gharial breeding program commenced in 1976, with three gharials in a large oval-shaped pool. The first successful breeding was completed in 1980, with 25 eggs laid. By 1985, the gharial population had reached a size of 206. From 1986 onward, 578 gharials were released into the Mahanadi River system, and 64 were transferred to other zoos. After the last adult male died in 2015, artificial incubation methods were employed to continue the breeding program, resulting in 71 healthy hatchlings (www.nandankan.org).

The induced breeding program resulted in an increase in the number of gharials to 206 individuals in 1985. After successful breeding, the hatchlings were released into natural water bodies. Between 1986–1987, a total of, 81 gharials were introduced into the Mahanadi River system.

In July 2019, a revised strategy for the recovery of gharials in Satkosia Gorge was formulated. Six transmitter-tagged gharials of both sexes from Nandankanan Biological Park were released into Satkosia Gorge to monitor their survival. Efforts have also been made to protect and restore the natural habitat of Satkosia Gorge to provide a conducive environment for crocodiles. The status of the Satkosia Gorge Crocodile Conservation Project has shown promising developments. As of early 2024, the population consists of 97 muggers and 14 gharials, which is an increase from 81 muggers and 7 gharials in the previous year. In 2019, a new initiative titled "The Mahanadi River Gharial Species Recovery Plan" was launched to conserve the gharials.

5.9 Conclusion

This study addresses various aspects of the natural history of the crocodilian fauna, which are crucial for understanding the vulnerabilities of crocodiles in the Mahanadi River. The Mahanadi River system, once a conducive habitat for all three crocodilian species, is now ecologically stressed and fragile, making it difficult to sustain crocodile populations. The factors driving the decline of crocodilian fauna include river flow modifications due to regulation, pollution, sand mining, and the use of fishing gear that harm crocodiles. Additionally, intrinsic characteristics such as long incubation periods and specific basking patterns further increase the vulnerability of crocodilian populations.

Several conservation measures have been implemented in the Mahanadi River system. These include the Nandankanan Wildlife Sanctuary's captive breeding program, the Bhitarkanika Crocodile Conservation Project, and the Satkosia Gorge conservation efforts, along with restocking programs. Moving forward, conservation strategies must address these challenges through habitat restoration, stricter enforcement of protected areas, community engagement, and sustainable river management practices. "The Mahanadi River Gharial Species Recovery Plan" is a positive step, incorporating lessons from previous efforts and adopting a more holistic approach to species recovery. The long-term survival of these reptiles in the Mahanadi River ecosystem will rely on continued collaboration among government agencies, conservation organizations, local communities, and researchers.

Acknowledgement The authors are thankful to the R&D Division, DoWR, RD & GR, Ministry of Jal Shakti, Government of India, for providing financial assistance (No. 10/28/2022-R&D/, July 26, 2023) for sponsoring the conference where this paper was presented.

Funding Statement No funding was received to conduct this study.

Declaration of Interest The authors declare that they have no known competing financial interests or personal relationships that could have appeared to influence the work reported in this paper.

References

Adhya S (2012) A review of crocodile conservation in Bhagabatpur West Bengal. INDEXED BY Impact Factor, 9

Basu SK, Das S, Cetzal-Ix W (2023) Indian saltwater crocodiles (Crocodylus porosus Schneider, 1801) and their conservation perspective. Annales Universitatis Paedagogicae Cracoviensis Studia Naturae 8(1). https://doi.org/10.24917/25438832.8.10

Behera BC, Mishra RR, Patra JK, Dutta SK, Thatoi HN (2014) Physico-chemical properties of water sample collected from mangrove ecosystem of Mahanadi River Delta, Odisha, India. Am J Mar Sci 2(1):19–24

Chaudhari D, Chaudhari A, Vaidh I, Patel H (2022) First confirmed record of mugger crocodile (*Crocodylus palustris*) from the Purna River, Tapi District, Gujarat. Reptiles Amphibians 29(1):436–438

Chavan UM, Borkar MR (2023) Observations on cooperative fishing, use of bait for hunting, propensity for marigold flowers and sentient behaviour in mugger crocodiles *Crocodylus palustris* (lesson, 1831) of river Savitri at Mahad, Maharashtra, India. J Threatened Taxa 15(8):23750–23762

Chen C, Holyoak M, Xu J et al (2023) Range restriction, climate variability, and human-related risks imperil lizards worldwide. Glob Ecol Biogeogr 32:780. https://doi.org/10.1111/geb.13655

Choudhary S, Choudhury BC, Gopi GV (2018) Spatio-temporal partitioning between two sympatric crocodilians (*Gavialis gangeticus* & *Crocodylus palustris*) in Katarniaghat wildlife sanctuary, India. Aquat Conserv Mar Freshwat Ecosyst 28(5):1067–1076

Da Silva A, Lenin J (2010) Mugger crocodile *Crocodylus palustris*. In: Manolis SC, Stevenson C (eds) Crocodiles. Status survey and conservation action plan, 3rd edn. Crocodile Specialist Group, pp 94–98

Daniel JC (2002) The book of Indian reptiles and amphibians. Oxford University Press

Das UK, Pradhan RK, Dash AP, Panth S (2021) Study on habitat preference and conservation of mugger crocodile (*Crocodylus palustris* lesson 1831) in Ghodahada dam of Ganjam district, Odisha, India. Int Educ Sci Res J 7(7):3–11

Debata S, Purohit S, Mahata A, Jena SK, Palita SK (2018) Mugger crocodile *Crocodylus palustris* lesson, 1831 (Reptilia: Crocodilia: Crocodylidae) in river Saberi of Godavari system in southern Odisha, India: conservation implications. J Threatened Taxa 10(6):11770–11774. https://doi.org/10.11609/jott.3974.10.6.11770-11774

Desai B, Bhowmik T, Srinivasan R, Whitaker N, Ghosal R (2024) Monitoring the stress physiology of free-ranging mugger crocodiles (*Crocodylus palustris*) across diverse habitats within Central Gujarat, India. Conserv Physiol 12(1)

Dutta SK, Nair MV, Mohapatra PP, Mahapatra AK (2009) Amphibians and reptiles

Ganesan N (2007) Gharial god and tiger goddess in the Indus Valley: some aspects of bronze age Indian religion. Museum of Fine Arts, Houston

Gopi GV, Pandav B (2009) Humans sharing space with Crocodylus porosus in Bhitarkanika wildlife sanctuary: conflicts and options. Curr Sci 96(4):459–460

Hazra S, Das S, Ghosh A, Raju PV, Patel A (2020) The Mahanadi delta: a rapidly developing delta in India. In: Deltas in the Anthropocene, pp 53–77. https://doi.org/10.1002/aqc.2911

Kar S (2013) Crocodile conservation programme in Odisha, India, with special reference to saltwater crocodiles, Crocodylus porosus of Bhitarkanika mangrove ecosystem. In: Crocodiles. Proceedings of the 22nd working meeting of the IUCN-SSC crocodile specialist group. IUCN, pp 290–294

Khan W, Hore U, Mukherjee S, Mallapur G (2020) Human-crocodile conflict and attitude of local communities toward crocodile conservation in Bhitarkanika wildlife sanctuary, Odisha, India. Mar Policy 121:104135. https://doi.org/10.1016/j.marpol.2020.104135

Kumar A, Kumar S, Zaidi YF, Kanaujia A (2012) A review on status and conservation of saltwater crocodile (*Crocodylus porosus*) in India. Int Day Biol Diver Mar Biodiver Rep 22:141–148

Lamichhane S, Bhattarai D, Karki JB, Gautam AP, Pandeya P, Tripathi S, Mahat N (2022) Population status, habitat occupancy and conservation threats to mugger crocodile (*Crocodylus palustris*) in Ghodaghodi lake complex, Nepal. Glob Ecol Conserv 33:e01977

Lloyd-Jones LR, Brien ML, Feutry P, Lawrence E, Beri P, Booth S, Coulson S, Baylis SM, Villiers K, Taplin LE, Westcott DA (2023) Implications of past and present genetic connectivity for management of the saltwater crocodile (*Crocodylus porosus*). Evol Appl 16(4):911–935. https://doi.org/10.1111/eva.13545

Marshall B, Strine C, Hughes A (2020) Thousands of reptile species threatened by underregulated global trade. Nat Commun 11:4738. https://doi.org/10.1038/s41467-020-18523-4

Mukhopadhyay A, Ghosh P, Chanda A, Ghosh A, Ghosh S, Das S, Hazra S (2018) Threats to coastal communities of Mahanadi delta due to imminent consequences of erosion–present and near future. Sci Total Environ 637:717–729

Murthy TSN, Pillai RS (1990) Crocodies In. In: Majupuria TC (ed) Wildlife wealth of India. Craftsman Press, Bangkok, pp 256–260. 667p

Nair T (2019) Crocodile conservation doesn't need a straw man. The Wire. https://science.thewire.in/environment/crocodile-conservation-doesnt-need-a-straw-man/

Nawab A (2013) Conservation prospects of smooth-coated otter Lutrogale perspicillata (Geoffroy saint-Hilaire, 1826) in Rajasthan. In: Faunal heritage of Rajasthan, India: conservation and management of vertebrates. Springer International Publishing, Cham, pp 273–283

Newman S (2023) Saltwater crocodile facts, habitat, bite, diet, life cycle, pictures. Animal Spot. https://www.animalspot.net/saltwater-crocodile.html

Panda UC, Sundaray SK, Rath P, Nayak BB, Bhatta D (2006) Application of factor and cluster analysis for characterization of river and estuarine water systems–a case study: Mahanadi River (India). J Hydrol 331(3–4):434–445

Panda AK, Katdare S, Gawan S, Sharma SP, Badola R, Hussain SA (2023) Population status and factors influencing the distribution of critically endangered gharial (*Gavialis gangeticus*) in a regulated unprotected river system in India. Glob Ecol Conserv 46:e02547. https://doi.org/10.1016/j.gecco.2023.e02547

Patil SR, Atigre RH, Patil SV (2012) First record of Mugger Crocodile Crocodylus Palustrius (Lesson, 1831) from River Kadavi at Sarud, Tal, Shahuwadi, Dist. Kolhapur, MS, India. In: International proceedings of chemical, biological & environmental engineering, p 47

Rafferty JP (2024, February 9) Mugger: size, attacks, tool use, & facts. Encyclopedia Britannica. https://www.britannica.com/animal/mugger

Rao RJ (1988) Nesting ecology of the gharial in National Chambal Sanctuary. Wildlife Institute of India

Rath LP, Dash SK, Khan A, Mohapatra RK, Kumar S, Nair MV, Paul S, Maharana S (2024) Monitoring the movement, habitat use and mortality of captive-bred reintroduced gharials in the Mahanadi River, India. Aquat Conserv 34(2). https://doi.org/10.1002/aqc.4097

Ratha KC (2019) Growing industrialization and its cumulative impacts: insights from the Mahanadi River Basin. Soc Change 49(3):519–530. https://doi.org/10.1177/0049085719863904

Sagar SR, Singh LAK (1993) Captive breeding and rehabilitation of mugger crocodile (*Crocodylus Palustris*) in Similipal Tiger Reserve. Indian Forester 119:807–807

Saikia PK (2013) Indian gharial (*Gavialis gangeticus*): status, ecology and conservation. In: Bentham science publishers eBooks, pp 76–100. https://doi.org/10.2174/97816080548551130100007

Sharma RK, Singh LAK (2015) Status of mugger crocodile (*Crocodylus palustris*) in National Chambal Sanctuary after thirty years and its implications on conservation of gharial (*Gavialis gangeticus*). Zoo's Print 30(5):9–16

Sharma SP, Ghazi MG, Katdare S, Dasgupta N, Mondol S, Gupta SK, Hussain SA (2021) Microsatellite analysis reveals low genetic diversity in managed populations of the critically endangered gharial (*Gavialis gangeticus*) in India. Sci Rep 11(1):5627

Singh LAK, Bustard HR (1977a) Studies on Indian gharial *Gavialis gangeticus* (Gmelin): V preliminary observation on maternal behavior. Indian For 103:671–678

Singh LAK, Bustard HR (1977b) Studies on the Indian gharial (*Gavialis gangeticus Gmelin*) (Reptilia, Crocodilia) V: preliminary observations on maternal behaviour. Indian Forester 103(10):671–678. https://doi.org/10.36808/if/1977/v103i10/11814

Singh LAK, Kar SK (2006) Status of the saltwater crocodile in Orissa: an overview. J Bombay Nat Hist Soc 103(2/3):274

Somanna K, Reddy TS, Rao MS (2016) Geomorphology and evolution of the modern Mahanadi delta using remote sensing data. Int J Sci Res 5(2):1329–1335

Sundaray SK, Panda UC, Nayak BB, Bhatta D (2006) Multivariate statistical techniques for the evaluation of spatial and temporal variations in water quality of the Mahanadi river–estuarine system (India)–a case study. Environ Geochem Health 28:317–330

Vashistha G, Mungi NA, Lang JW, Ranjan V, Dhakate PM, Khudsar FA, Kothamasi D (2021) Gharial nesting in a reservoir is limited by reduced river flow and by increased bank vegetation. Sci Rep 11:4805. https://doi.org/10.1038/s41598-021-84143-7

Whitaker R (1987) The management of crocodilians in India. In: Webb GJW, Manolis SC, Whitehead PJ (eds) Wildlife management: crocodiles and alligators, pp 63–72

Whitaker R, Basu D (1981) The gharial *Gavialis gangeticus*: a review. J Bombay Nat Hist Soc 79:531–548

Whitaker R, Whitaker Z (1989) Ecology of mugger crocodile. In: Crocodiles: their ecology, management, and conservation. International Union for Conservation of Nature, pp 276–297

Williamson SA, Evans RG, Manolis SC, Webb GJ, Reina RD (2017) Ecological and evolutionary significance of a lack of capacity for extended developmental arrest in crocodilian eggs. R Soc Open Sci 4(12):171439. https://doi.org/10.1098/rsos.171439

Diversity and Distribution of Ichthyofauna in the Mahanadi River Basin in India

6

Snigdha Snehanjali and Jay Prakash Bhatt

Abstract

The Mahanadi River holds significant ecological importance because of its large basin area of 141,600 sq. km. This basin has high potential for water resources, including rivers, lakes, reservoirs, and salt lagoons, which are often inhabited by diverse fish fauna. This study aims to investigate the diversity and distribution patterns of freshwater fish, the factors affecting fish diversity, their conservation status, and the threats to fish fauna. In total, 176 species were confirmed from various sources. The Michaelis–Menten equation can be used to predict approximately 209 species that are expected to inhabit the Mahanadi River Basin. These species are grouped into 119 genera across 59 families. The fish fauna includes one critically endangered species, five endangered species, and seven vulnerable species. The fish species exhibited a hump-shaped distribution pattern along both latitudinal and longitudinal gradients, with relatively larger areas in the mid-latitudes and mid-longitudes driving this distribution. Major threats to the fish fauna of the Mahanadi River Basin include flow modifications due to river regulation, pollution, indiscriminate fishing, and the unavailability of sufficient natural food. Given this, a substantial conservation strategy for the fish fauna of the Mahanadi River basin is warranted. Fishery scientists suggest several conservation measures, including the establishment of hatcheries for threatened species, improvements in recruitment processes, the regulation of mesh sizes in fishing nets, fish ranching, and raising awareness among fishing communities.

S. Snehanjali · J. P. Bhatt (✉)
Faculty of Science, Sri Sri University, Cuttack, Odisha, India
e-mail: jay.b@srisriuniversity.edu.in

J. P. Bhatt, A. K. Mishra (eds.), *Mahanadi River*,
https://doi.org/10.1007/978-981-96-1048-8_6

Keywords

Mahanadi River · Ichthyofaunal species · Diversity · Conservation status · Geographical distribution

6.1 Introduction

Mahanadi is one of the largest rivers in peninsular India and is spread primarily over the states of Chhattisgarh and Odisha (Singh and Das 2018). It drains an area of 1,41,589 sq. km., with a total length of 851 km from its origin to outfall into the sea. The land use/land cover of the Mahanadi Basin essentially includes agriculture (54.27%), forests (32.74%), wasteland (5.24%), water bodies (4.45%), and built-up areas (3.3%) (India-WRIS 2012), indicating prosperous ecosystem services. Rich ichthyofaunal diversity and fisheries constitute one of the major ecosystem services provided by the Mahanadi River system (Sajina et al. 2024). The livelihood of a reasonable size of the population in the states Odisha and Chhattisgarh depends largely on the fishery resources of the Mahanadi Basin. The basin also has high potential for the development of fisheries (Lall 2020; Tandi and Tandi 2018).

Like other rivers in India, the Mahanadi River is under severe stresses, such as sewage outfall, industrial effluents, agricultural runoff, river regulation, etc. (Mishra and Nayak 2014; India-WRIS 2012). All these stressors are well known to affect fish and fisheries adversely, which could then lead to habitat loss, pollution, flow modification, habitat fragmentation, the spread of invasive species, etc. (Mishra and Nayak 2014; Singh and Das 2018). Fish are considered good responders to alterations in their physical and chemical characteristics (Dash et al. 2018). The aspects of fish and fisheries in the Mahanadi Basin have been addressed by various authors; however, the majority of these contributions revolves around the fish inventory, taxonomy, and fisheries (Nandini and Parida 2020). Studies on the biogeographic distribution of fish along longitudinal and latitudinal gradients, drivers of distribution, and conservation aspects are largely overlooked. Thus, present study is an effort to fill these gaps. This study contributes to our understanding of the environmental pressures faced by Mahanadi fishes and their conservation and management.

6.2 Materials and Methods

The Mahanadi River Basin stretches between 19° 8′N and 23° 32′N latitude and 80° 28′E and 86° 43′ E longitude (Fig. 6.1). The important tributaries and distributaries delineating the subbasins are the Seonath, Hasdeo, Tel, Ib, Mand, Ong, Jonk, Kathajodi, Birupa, Chitropala and Paika rivers. The river basin is bordered to the north by the Ganga and Narmada river basins; to the east by the Baitarani and Brahmani river basins; and to the northwest, west, and southwest by the Godavari Basin. The Mahanadi Basin bordered south and southwest by small river basins located between the Mahanadi and Godavari basins. It finally opens to the Bay of Bengal.

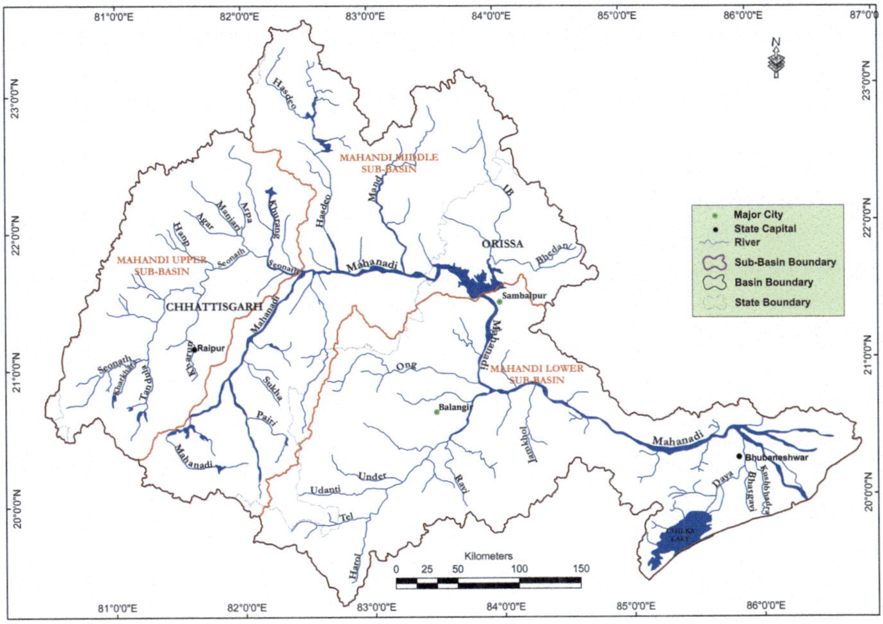

Fig. 6.1 Basin area of the Mahanadi River

6.3 Fish Species Richness

The fish species data were sourced from previously published data, online data sets, and government records (Singh 2014; Dash et al. 2018; Baliarsingh et al. 2020; Rejani et al. 2023; Singh et al. 2006, 2015, 2020; Bharati et al. 2019; Choubey and Qureshi 2013; Niyazi et al. 2020; ; Tyagi et al. 2021; Patel et al. 2016a, b, c; Baliarsingh and Kosygin 2017; Kand et al. 2023; Gopi et al. 2018; Panda and Sahoo 2020; Tamboli and Jha 2012; Baliarsingh et al. 2017; Johnson et al. 2023; Khanani and Aggarwal 2018). Each species was subjected to the International Union for Conservation of Nature (IUCN) Red List (www.iucnredist.org/) to assess its conservation status. In addition to the published data, we consulted other sources, such as a fish base (www.fishbase.org), to gather other information, such as endemism and habitat.

6.4 Distribution Pattern

The Mahanadi River Basin was divided into latitudinal and longitudinal gradients ranging from 19°N to 23°N and from 80°E to 86°E, respectively. The latitudinal and longitudinal gradients were then divided into five and seven bands, respectively, with an interval of 1 degree. We compared total species richness against these gradients via published data. We followed the interpolation method to estimate the number of fish species in each latitudinal and longitudinal band (see Bhatt et al. 2012). We have also marked the lowest and highest latitudinal and longitudinal records for each species and assumed that the species was found in all bands between its lowest and highest latitudinal or longitudinal records.

6.5 Expected Species Richness

We assume that the Mahanadi River has not been fully explored and that there are fair possibilities for more species. To find the expected number of fish species, we used the Michaelis–Menten equation (Dahanukar et al. 2003):

$$S = \frac{S \max N}{Km + N}$$

where S is the number of species, N is the sampling unit, Smax is the maximum number of species that could be present, and km is the Michaelis–Menten constant.

6.6 Results

6.6.1 Species Diversity

In total, 176 species were confirmed from the freshwater sources of the Mahanadi River with the help of secondary literature. These species were grouped into 119 genera across 59 families (Table 6.1). The Cyprinidae family was the largest, accounting for 26.2% of the total species. The Cyprinidae family was followed by the families Danionidae (8.3%), Bagridae (5.5%), Channidae (3.9%), and Dorosomatidae (3.9%). In total, 24 species (13.4%) shared their habitat with seawater environments and were diadromous in their migratory habits.

6.6.2 Conservation Status

The majority of the fish species (79.3%) in the freshwater of the Mahanadi River Basin was classified as "least concern," according to the IUCN Red List. A single "critically endangered" species recorded was *Epalzeorhynchos bicolor*. The "endangered" category of the IUCN Red List included five species, viz., *Hypselobarbus curmuca*, *Pangasianodon hypophthalmus*, *Pangasius sutchi*, *Puntius vittatus*, and *Tor putitora*. The "vulnerable" category of threatened species included seven species, namely, *Bagarius bagarius*, *Channa orientalis*, *Cyprinus carpio*, *Wallago attu*, *Oreochromis mossambicus*, *Cirrhinus cirrhosus*, and *Bagarius yarrelli*. The remaining species were classified as "near-threatened" or "data deficient." In total, seven species were not evaluated for their conservation status (Table 6.1).

Table 6.1 List of fish species in the Mahanadi River Basin along with conservation status and habitat. (Fishery- O= ornamental fish, F= fish food; Conservation status – NE = not evaluated, LC = least concern, DD = data deficient' VU = vulnerable, EN = endangered, CR = critically endangered; Habitat - F = freshwater M = marine)

S.No.	Family	Scientific name	Common name	Fishery	Conservation status	Habitat
1	Lobotidae	*Datnioides polota* (Hamilton, 1822)	Silver tiger perch	O	LC	F
2	Sciaenidae	*Johnius dussumieri* (Cuvier, 1830)	Sin croaker	F	LC	F, M
3	Ephippidae	*Platax pinnatus* (Linnaeus, 1758)	Dusky batfish	O	NE	F
4	Anabantidae	*Anabas oligolepis* (Hamilton, 1822)	Gangetic koi	O	DD	F
5	Anabantidae	*Anabas testudineus* (Bloch, 1792)	Climbing perch	F/O	LC	F
6	Badidae	*Badis badis* (Hamilton, 1822)	Badis	O	LC	F, M
7	Channidae	*Channa gachua* (Hamilton, 1822)	Red seam snakehead	F	LC	F
8	Channidae	*Channa marulius* (Hamilton, 1822)	Great snakehead	F	LC	F
9	Channidae	*Channa orientalis* (Bloch & Schneider, 1801)	Walking snakehead	O	VU	F
10	Channidae	*Channa striata* (Bloch, 1793)	Striped snakehead	F	LC	F
11	Anabantidae	*Anabas cobojius* (Hamilton, 1822)	Gangetic koi	O	DD	F
12	Channidae	*Channa punctata* (Bloch, 1793)	Spotted snakehead	F	LC	F
13	Osphronemidae	*Trichogaster lalius* (Hamilton, 1822)	Dwarf gourami	F	LC	F
14	Nandidae	*Nandus nandus* (Hamilton, 1822)	Gangetic leaf fish	O	LC	F
15	Pristolepididae	*Pristolepis fasciata* (Bleeker, 1851)	Malayan leaf fish	F/O	LC	F
16	Osphronemidae	*Trichogaster fasciata* (Bloch & Schneider, 1801)	Banded gourami	F	LC	F
17	Anguillidae	*Anguilla bengalensis* (Gray, 1831)	Indian mottled eel	F	NT	F,M
18	Hemiramphidae	*Hyporhamphus limbatus* (Valenciennes, 1847)	Congaturi half-beak	F	LC	F, M
19	Belonidae	*Strongylura strongylura* (van Hasselt, 1823)	Spottail needlefish	O	NE	F
20	Belonidae	*Xenentodon cancila* (Hamilton, 1822)	Garfish	F/O	LC	F
21	Carangidae	*Caranx sexfasciatus* (Quoy & Gaimard, 1825)	Bigeye trevally	O	LC	F,M
22	Terapontidae	*Terapon jarbua* (Fabricius, 1775)	Jarbua terapon	F/O	LC	F
23	Protopteridae	*Protopterus amphibius* (Peters, 1844)	Gilled lungfish	F	LC	F

(continued)

Table 6.1 (continued)

S.No.	Family	Scientific name	Common name	Fishery	Conservation status	Habitat
24	Cichlidae	*Oreochromis mossambicus* (Peters, 1852)	Mozambique tilapia	F	VU	F
25	Cichlidae	*Oreochromis niloticus* (Linnaeus, 1758)	Nile tilapia	F	LC	F
26	Dorosomatidae	*Anodontostoma chacunda* (Hamilton, 1822)	Chacunda gizzard shad	F	LC	F,M
27	Dorosomatidae	*Gudusia chapra* (Hamilton, 1822)	Indian river shad	F	LC	F
28	Pristigasteridae	*Pellona ditchela* (Valenciennes, 1847)	Indian pellona	F	LC	F, M
29	Dorosomatidae	*Sardinella fimbriata* (Valenciennes, 1847)	Fringescale sardinella	F	LC	F
30	Engraulidae	*Stolephorus commersonii* (Lacepède, 1803)	Commerson's anchovy	F	LC	F
31	Dorosomatidae	*Amblygaster leiogaster* (Valenciennes, 1847)	Smoothbelly sardinella	F/O	LC	M
32	Engraulidae	*Thryssa setirostris* (Broussonet, 1782)	Longjaw thryssa	O	LC	F, M
33	Ehiravidae	*Corica soborna* (Hamilton, 1822)	Ganges River sprat	F	LC	F, M
34	Dorosomatidae	*Gonialosa manmina* (Hamilton, 1822)	Ganges River gizzard shad	F/O	LC	F
35	Dorosomatidae	*Hilsa kelee* (Cuvier, 1829)	Kelee shad	F	LC	F, M
36	Dorosomatidae	*Tenualosa ilisha* (Hamilton, 1822)	Hilsa shad	F	LC	F
37	Nemacheilidae	*Acanthocobitis botia* (Hamilton, 1822)	Mottled loach	O	LC	F
38	Danionidae	*Amblypharyngodon mola* (Hamilton, 1822)	Mola carplet	F	LC	F
39	Cyprinidae	*Bangana dero* (Hamilton, 1822)	Kalabans	F	LC	F
40	Cyprinidae	*Bangana dero* (Hamilton, 1822)	Kalabans	F	LC	F
41	Danionidae	*Barilius barila* (Hamilton, 1822)	Barred baril	O	LC	F
42	Danionidae	*Barilius bendelisis* (Hamilton, 1807)	Indian Hill Trout	O	LC	F
43	Cyprinidae	*Barilius vagra* (Hamilton, 1822)	Vagra baril	F/O	LC	F
44	Danionidae	*Cabdio morar* (Hamilton, 1822)	Morari	F/O	LC	F
45	Cyprinidae	*Carassius auratus* (Linnaeus, 1758)	Goldfish	O	LC	F
46	Cyprinidae	*Cirrhinus cirrhosus* (Bloch, 1795)	Mrigal carp	F	VU	F
47	Cyprinidae	*Cirrhinus reba* (Hamilton, 1822)	Reba carp	F	LC	F
48	Cyprinidae	*Cirrhinus mrigala* (Hamilton, 1822)	Mrigal carp	F	LC	F
49	Cyprinidae	*Cyprinus carpio* (Linnaeus, 1758)	Common carp	F	VU	F

50	Danionidae	*Danio rerio* (Hamilton, 1822)	Zebra danio	O	LC	F
51	Cyprinidae	*Esomus danrica* (Hamilton, 1822)	Indian flying barb	O	LC	F
52	Cyprinidae	*Garra annandalei* (Hora, 1921)	Tunga garra	O	LC	F
53	Cyprinidae	*Garra gotyla* (Gray, 1830)	Sucker head	O	LC	F
54	Cyprinidae	*Garra mullya* (Sykes, 1839)	Sucker fish	O	LC	F
55	Cyprinidae	*Gibelion catla* (Hamilton, 1822)	Catla	F	LC	F
56	Xenocyprididae	*Hypophthalmichthys molitrix* (Valenciennes, 1844)	Silver carp	F	NT	F
57	Cyprinidae	*Hypselobarbus curmuca* (Hamilton, 1807)	Curmuca barb	F	EN	F
58	Cyprinidae	*Hypselobarbus kolus* (Sykes, 1839)	Kolus	O	LC	F
59	Nemacheilidae	*Indoreonectes evezardi* (Day, 1872)	Murangi	O	LC	F
60	Cyprinidae	*Labeo angra* (Hamilton, 1822)	Angra labeo	F	LC	F
61	Cyprinidae	*Labeo bata* (Hamilton, 1822)	Bata	F	LC	F
62	Cyprinidae	*Labeo boga* (Hamilton, 1822)	Boga labeo	F	LC	F
63	Cyprinidae	*Labeo boggut* (Sykes, 1839)	Boggut labeo	F	LC	F
64	Cyprinidae	*Labeo calbasu* (Hamilton, 1822)	Orangefin labeo	F	LC	F
65	Cyprinidae	*Labeo catla* (Hamilton, 1822)	Catla	F	LC	F
66	Cyprinidae	*Labeo dyocheilus* (McClelland, 1839)	Boalla Kali	F	LC	F
67	Cyprinidae	*Labeo fimbriatus* (Bloch, 1795)	Fighting-lipped peninsula carp	F	LC	F
68	Cyprinidae	*Labeo gonius* (Hamilton, 1822)	Kuria labeo	F	LC	F
69	Cyprinidae	*Labeo pangusia* (Hamilton, 1822)	Pangasia Labeo	F	LC	F
70	Cyprinidae	*Labeo rohita* (Hamilton, 1822)	Rohu	F	LC	F
71	Danionidae	*Laubuka laubuca* (Hamilton, 1822)	Indian glass barb	O	LC	F
72	Cobitidae	*Lepidocephalichthys guntea* (Hamilton, 1822)	Guntea loach	O	LC	F
73	Danionidae	*Opsarius barna* (Hamilton, 1822)	Barna baril	F	LC	F

(continued)

Table 6.1 (continued)

S.No.	Family	Scientific name	Common name	Fishery	Conservation status	Habitat
74	Cyprinidae	*Osteobrama cotio* (Hamilton, 1822)	Mou	O	LC	F
75	Danionidae	*Oxygaster bacaila* (Hamilton, 1822)	Large razorbelly minnow	F/O	LC	F
76	Cyprinidae	*Parapsilorhynchus swaini* (Hora,1921)	Mahanadi minnow	O	LC	F
77	Cyprinidae	*Parapsilorhynchus odishaensis* (Baliarsingh, Kosygin & Swain, 2017)	F	DD	F	
78	Cyprinidae	*Pethia conchonius* (Hamilton, 1822)	Rosy barb	O	LC	F
79	Cyprinidae	*Pethia phutunio* (Hamilton, 1822)	Spottedsail barb	O	LC	F
80	Cyprinidae	*Pethia ticto* (Hamilton, 1822)	Ticto barb	O	LC	F
81	Cyprinidae	*Puntius chola* (Hamilton, 1822)	Swamp barb	O	LC	F
82	Cyprinidae	*Puntius dorsalis* (Jerdon, 1849)	Long snouted barb	F/O	LC	F
83	Cyprinidae	*Puntius sophore* (Hamilton, 1822)	Pool barb	O	LC	F
84	Cyprinidae	*Puntius terio* (Hamilton, 1822)	Onespot barb	O	LC	F
85	Danionidae	*Rasbora daniconius* (Hamilton, 1822)	Slender rasbora	O	LC	F
86	Danionidae	*Rasbora einthovenii* (Bleeker, 1851)	Brilliant rasbora	O	LC	F
87	Danionidae	*Salmostoma phulo* (Hamilton, 1822)	Finescale razorbelly minnow	F	LC	F
88	Nemacheilidae	*Schistura denisoni* (Day, 1867)	Aiyra	O	LC	F
89	Danionidae	*Securicula gora* (Hamilton, 1822)	Gora chela	F/O	LC	F
90	Cyprinidae	*Systomus sarana* (Hamilton, 1822)	Olive barb	F	LC	F
91	Cyprinidae	*Tor mahanadicus* (David, 1953)	Mahanadi mahseer	F	LC	F
92	Cyprinidae	*Tor putitora* (Hamilton, 1822)	Putitor mahseer	F/O	EN	F
93	Xenocyprididae	*Ctenopharyngodon idella* (Valenciennes, 1844)	Grass carp	F	LC	F
94	Cyprinidae	*Epalzeorhynchos bicolor* (Smith, 1931)	Redtail sharkminnow	O	CR	F
95	Xenocyprididae	*Hypophthalmichthys nobilis* (Richardson, 1845)	Bighead carp	F	DD	F
96	Cyprinidae	*Puntius vittatus* (Day 1865)	Greenstripe barb	F	EN	F
97	Danionidae	*Barilius barna* (Hamilton, 1822)	Dudhnea	O	LC	F

98	Danionidae	*Devario aequipinnatus* (McClelland, 1839)	Giant danio	O	LC	F
99	Danionidae	*Devario devario* (Hamilton, 1822)	Sind danio	O	LC	F
100	Cyprinidae	*Oreichthys cosuatis* (Hamilton, 1822)	Indian hifin barb	O	LC	F
101	Cyprinidae	*Osteobrama vigorsii* (Sykes, 1839)	Bheema osteobrama	F	LC	F
102	Cyprinidae	*Pethia gelius* (Hamilton, 1822)	Golden barb	O	LC	F
103	Danionidae	*Salmostoma bacaila* (Hamilton, 1822)	Large razorbelly minnow	F	LC	F
104	Cyprinidae	*Tor tor* (Hamilton, 1822)	Tor barb	F	DD	F
105	Aplocheilidae	*Aplocheilus panchax* (Hamilton, 1822)	Blue panchax	F/O	LC	F
106	Poeciliidae	*Gambusia affinis* (Baird & Girard, 1853)	Mosquito fish	O	LC	F
107	Megalopidae	*Megalops cyprinoides* (Broussonet, 1782)	Indo-Pacific tarpon	F	DD	F, M
108	Gobiidae	*Glossogobius giuris* (Hamilton, 1822)	Tank goby	F	LC	F, M
109	Chanidae	*Chanos chanos* (Fabricius, 1775)	Milkfish	F	LC	F, M
110	Mugilidae	*Mugil cephalus* (Linnaeus, 1758)	Flathead grey mullet	F	LC	F, M
111	Mugilidae	*Planiliza parsia (Hamilton, 1822)*	Goldspot mullet	F	NE	F, M
112	Mugilidae	*Planiliza tade (Fabricius, 1775)*	Tade gray mullet	F	DD	F, M
113	Mugilidae	*Rhinomugil corsula (Ham.-Buch.)*	Corsula	F	LC	F
114	Mugilidae	*Osteomugil cunnesius (Valenciennes, 1836)*	Longarm mullet	F	NE	F, M
115	Mugilidae	*Planiliza planiceps (Valenciennes, 1836)*	Stripped mullet	F	LC	F, M
116	Mullidae	*Upeneus sulphureus (Cuvier, 1829)*	Sulphur goatfish	F	LC	F, M
117	Notopteridae	*Chitala chitala* (Hamilton, 1822)	Clown knifefish	O	NT	F
118	Notopteridae	*Notopterus notopterus (Pallas, 1769)*	Bronze featherback	F/O	LC	F
119	Ambassidae	*Ambassis gymnocephalus* (Lacepède, 1802)	Bald glassy	F	LC	F,M
120	Ambassidae	*Chanda nama* (Hamilton, 1822)	Elongate glass-perchlet	O	LC	F
121	Polynemidae	*Eleutheronema tetradactylum* (Shaw, 1804)	Fourfinger threadfin	F	NE	F, M
122	Gerreidae	*Gerres setifer* (Hamilton, 1822)	Small Bengal silver biddy	F	NE	F, M

(continued)

Table 6.1 (continued)

S.No.	Family	Scientific name	Common name	Fishery	Conservation status	Habitat
123	Sciaenidae	*Johnius coitor* (Hamilton, 1822)	Coitor croaker	F	LC	F, M
124	Latidae	*Lates calcarifer* (Bloch, 1790)	Barramundi	F/O	LC	F
125	Ambassidae	*Parambassis lala* (Hamilton, 1822)	Highfin glassy perchlet	O	NT	F
126	Ambassidae	*Parambassis ranga* (Hamilton, 1822)	Indian glassy fish	F	LC	F
127	Cynoglossidae	*Cynoglossus puncticeps* (Richardson, 1846)	Speckled tongue sole	O	LC	F, M
128	Trichiuridae	*Lepturacanthus savala* (Cuvier, 1829)	Savalai hairtail	O	NE	F, M
129	Scombridae	*Rastrelliger kanagurta* (Cuvier, 1816)	Indian mackerel	F	LC	F, M
130	Platycephalidae	*Platycephalus indicus* (Linnaeus, 1758)	Bartail flathead	F	DD	F
131	Ailiidae	*Ailia coila* (Hamilton, 1822)	Genetic ailia	F	NT	F
132	Ariidae	*Arius arius* (Hamilton, 1822)	Thread in sea catfish	F	LC	M
133	Sisoridae	*Bagarius bagarius* (Hamilton, 1822)	Goonch	F/O	VU	F
134	Sisoridae	*Bagarius yarrelli* (Hamilton, 1822)	Goonch catfish	F	VU	F
135	Chacidae	*Chaca chaca* (Hamilton, 1822)	Squarehead catfish	O	LC	F
136	Clariidae	*Clarias batrachus* (Linnaeus, 1758)	Philippine catfish	F/O	LC	F
137	Clariidae	*Clarias gariepinus* (Burchell, 1822)	North African catfish	F	LC	F
138	Clariidae	*Clarias magur* (Hamilton, 1822)	Magur	F	LC	F
139	Ailiidae	*Clupisoma garua* (Hamilton, 1822)	Garua bachcha	F	LC	F
140	Sisoridae	*Erethistes hara* (Hamilton, 1822)	Hara hara	F	LC	F
141	Schilbeidae	*Eutropiichthys vacha* (Hamilton, 1822)	Batchwa vacha	F	LC	F
142	Heteropneustidae	*Heteropneustes fossilis* (Bloch, 1794)	Stinging catfish	F/O	LC	F
143	Bagridae	*Mystus bleekeri* (Day 1877)	Day's mystus	O	LC	F
144	Bagridae	*Mystus cavasius* (Hamilton, 1822)	Gangetic mystus	F	LC	F
145	Siluridae	*Ompok bimaculatus* (Bloch, 1794)	Butter catfish	F/O	NT	F
146	Siluridae	*Ompok pabo* (Hamilton, 1822)	Pabo cat fish	F	NT	F
147	Pangasiidae	*Pangasianodon hypophthalmus* (Sauvage, 1878)	Striped catfish	F	EN	F
148	Pangasiidae	*Pangasius pangasius* (Hamilton, 1822)	Pangas catfish	F	LC	F
149	Pangasiidae	*Pangasius sutchi* (Fowler, 1937)	Iridescent shark	F	EN	F
150	Bagridae	*Sperata aorella* (Blyth, 1858)	Bagrid Catfish	F	LC	F

151	Siluridae	*Wallago attu* (Bloch & Schneider, 1801)	Wallago	F	VU	F
152	Amblycipitidae	*Amblyceps mangois* (Hamilton, 1822)	Indian torrent catfish	F/O	LC	F
153	Schilbeidae	*Eutropiichthys murius* (Hamilton, 1822)		F	LC	F
154	Bagridae	*Mystus gulio* (Hamilton, 1822)	Long whisker catfish	F	LC	F
155	Bagridae	*Rita chrysea* (Day, 1877)	Mahanadi rita	F	LC	F
156	Sisoridae	*Gagata cenia* (Hamilton, 1822)	Indian gagata	F	LC	F
157	Sisoridae	*Gagata gagata* (Hamilton, 1822)	Gangetic gagata	F	LC	F
158	Sisoridae	*Hara hara* (Hamilton, 1822)	Koshi Hara	F/O	LC	F
159	Bagridae	*Mystus tengara* (Hamilton, 1822)	Tengara catfish	O	LC	F
160	Bagridae	*Mystus vittatus* (Bloch, 1794)	Striped dwarf catfish	O	LC	F
161	Siluridae	*Ompok pabda* (Hamilton, 1822)	Pabdah cat fish	F	NT	F
162	Schilbeidae	*Pachypterus atherinoides* (Bloch, 1794)	Indian potasi	F/O	LC	F
163	Schilbeidae	*Pseudeotropis atherinoedes* (Bloch, 1794)	Indian potasi	F/O	LC	F
164	Bagridae	*Rita rita* (Hamilton, 1822)	Rita	F	LC	F
165	Schilbeidae	*Silonia silondia* (Hamilton, 1822)	Silond catfish	O	LC	F
166	Bagridae	*Sperata aor* (Hamilton, 1822)	Long whiskered cat fish	F	LC	F
167	Bagridae	*Sperata seenghala* (Sykes, 1839)	Giant river catfish	F	LC	F
168	Sillaginidae	*Sillago sihama* (Fabricius, 1775)	Silver sillago	F	LC	F, M
169	Mastacembelidae	*Mastacembelus armatus* (Lacepède, 1800)	Zig-zag eel	F	LC	F
170	Mastacembelidae	*Mastacembelus pancalus* (Hamilton, 1822)	Barred spiny eel	F	LC	F
171	Mastacembelidae	*Macrognathus aculeatus* (Bloch, 1786)	Lesser spiny eel	F/O	LC	F
172	Mastacembelidae	*Macrognathus aral* (Bloch & Schneider, 1801)	One stripe spiny eel	F	LC	F
173	Mastacembelidae	*Macrognathus pancalus* (Hamilton, 1822)	Barred spiny eel	O	LC	F
174	Synbranchidae	Monopterus cuchia (Hamilton, 1822)	Cuchia	F	LC	F
175	Sciaenidae	*Johnius gangeticus* (Talwar, 1991)		F	DD	F
176	Tetraodontidae	*Leiodon cutcutia* (Hamilton, 1822)	Ocellated pufferfish	O	LC	F

6.6.3 Distribution Pattern

The total species richness exhibited hump-shaped patterns along the latitudinal and longitudinal gradients of the Mahanadi River Basin. The maximum richness was observed at mid-latitudes and mid-longitudes. We did not find definite patterns of threatened species along the latitudinal and longitudinal gradients of the Mahanadi River Basin (Figs. 6.2 and 6.3).

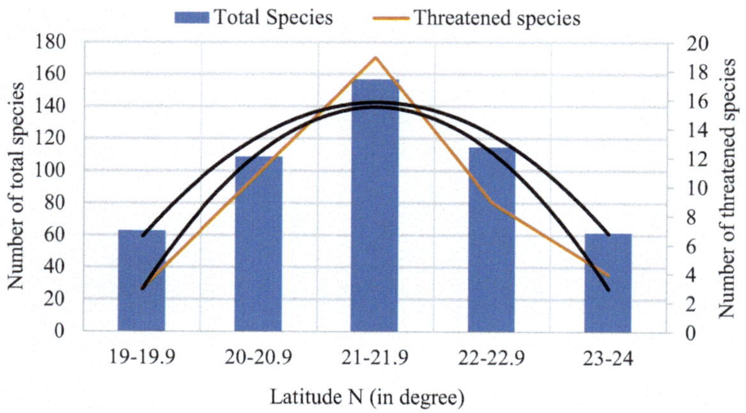

Fig. 6.2 Distribution patterns of the total fish species and threatened fish species in the Mahanadi River Basin along the latitudinal gradient. For threatened species, only near-threatened, vulnerable, endangered, and critically endangered species were considered for the assessment

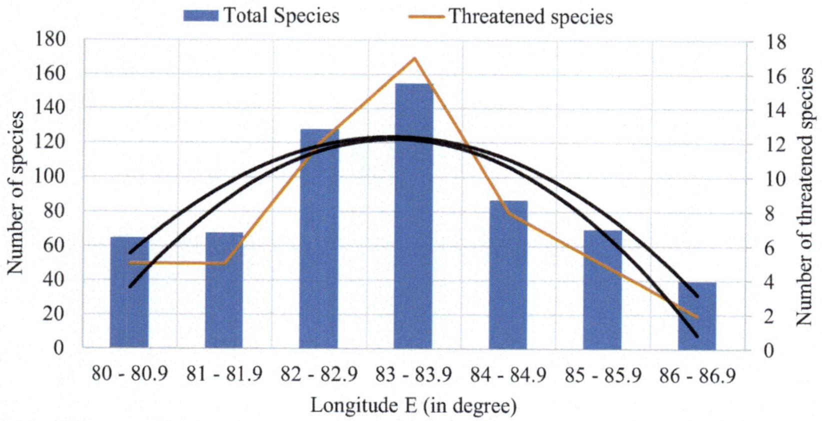

Fig. 6.3 Distribution patterns of the total fish species and threatened fish species in the Mahanadi River Basin along the longitudinal gradient. For threatened species, only near-threatened, vulnerable, endangered, and critically endangered species were considered for the assessment

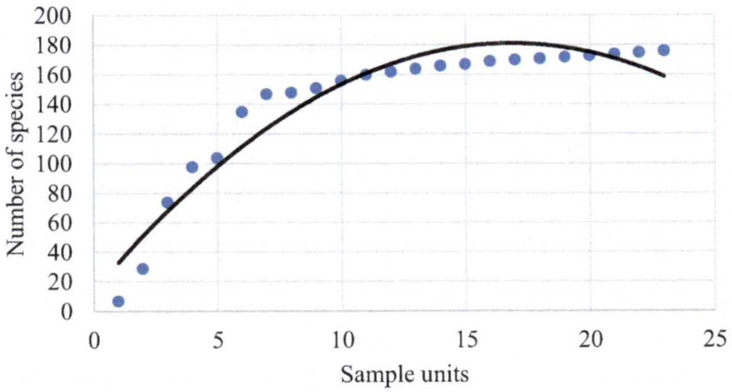

Fig. 6.4 The cumulative frequency of the fish species was determined, and the Michelis–Menten equation was used to find the expected number of fish species found in the Mahanadi River Basin ($S = 176$, km = 3.9, Smax = 209)

6.6.4 Expected Number of Species

The Michaelis–Menten equation was used to calculate the 209 species that are expected to inhibit the Mahanadi River (Fig. 6.4).

6.6.5 Fisheries

Approximately 99 species were reported to be of fisheries importance, while 53 species were ornamental.

6.7 Discussion

The Mahanadi River Basin is one of the largest river basins in India, accounting for nearly 4.28% of the country's total geographic area (Rejani et al. 2023). The size of the basin may be one of the causes of relatively high richness and diversity. Additionally, high richness can be attributed to other various basin characteristics, such as geomorphology, water quality, water flow, temperature variation, and rainfall (Spurgeon et al. 2018; Swain 2014; Pradhan et al. 2024). Additionally, the presence of the inland Chilika Lagoon contributes to the rich freshwater fish diversity in the Mahanadi Basin, as many fish species from the Chilika estuary share their habitat with the Mahanadi River and its tributaries (Hora 1940). While the current characteristics of this basin appear favorable for freshwater fish resources, it seems challenging to maintain these features in the future because of anthropogenic activities, particularly regarding water quality and river flow. The anthropogenic stresses on the freshwater fish resources of the Mahanadi River are evident (Tyagi et al. 2020).

Reports indicate that the Mahanadi River Basin once had abundant fishery resources, which are now in gradual decline due to various stresses, such as siltation, flow modification, overexploitation, insufficient natural food, and indiscriminate fishing (Ratha 2019; Tyagi et al. 2020; Singh et al. 2020; Moraga et al. 2022; Díaz et al. 2023). Notably, in total, 14 threatened species (vulnerable, endangered, critically endangered) have been reported in the Mahanadi River. Among these, species such as *Tor putitora*, *Bagarius bagarius*, and *Anguilla bengalensis* are significantly affected by human-induced activities (Tyagi et al. 2020).

The distribution patterns of fish species are largely affected by variations in area, climatic conditions, and other biogeographic features along latitudinal and longitudinal gradients at regional and global scales (Oberdorff et al. 1995, 2011). Although the Mahanadi Basin covers a large area, it spans only 5 degrees in latitude and 7 degrees in longitude, with approximately 99% of the basin lying within the tropical ecoregion (see India-WRIS 2012). Therefore, ecoregional differences in environmental variables cannot be considered the driving factors of distribution patterns. The only factor that seems to shape the distribution pattern of fish is area. The map of the Mahanadi Basin shows relatively high areas at mid-latitudes and mid-longitudes, where the richness of fish species is high. Research indicates that area plays a significant role in shaping species richness (Oberdorff et al. 2011). However, in the case of the Mahanadi River Basin, more studies are needed.

Information on diversity, distribution patterns, conservation status, and threats is essential for formulating conservation strategies (Li et al. 2012). We did not find substantial conservation strategies for fish in the Mahanadi River Basin. However, many fishery scientists have advocated various fish conservation measures, such as establishing specific fish habitats, preventing overexploitation, improving recruitment processes by banning fine-meshed nets, regulating the size of drag and gill nets, developing hatcheries for native threatened species for restocking purposes, fish ranching, raising awareness among fishing communities about the adverse effects of wanton fishing, banning the introduction of exotic species, and strengthening fishery policies (Singh et al. 2020; Tyagi et al. 2020; Bhatt and Manish 2023). Sustenance fisheries contribute to the economy of the Mahanadi Basin. Therefore, conservation planning cannot be formulated without involving fishermen communities and addressing their concerns. Additionally, there is a need to identify all other stakeholders that are directly and indirectly associated with the Mahanadi River Basin.

6.8 Conclusion

The Mahanadi River Basin supports rich and diverse freshwater fish fauna. The fish contribute to the sustenance fisheries of the basin, and the livelihood of a sizable population in the region depends on these fishery resources. Over the past few decades, fishery resources have been under stress due to human-induced activities such as river regulation, pollution, sand mining, and indiscriminate fishing. Considering all such threats to the fish of the Mahanadi Basin, a conservation strategy needs to be implemented. This strategy should involve all important stakeholders, including fishing communities, river regulatory authorities, governmental agencies of all concerned states, municipal corporations, and scientists from research institutes. Importantly, the Mahanadi Basin requires further exploration. The limited secondary literature on the fish and fisheries of the Mahanadi River indicates that more research is needed; therefore, more fish species from the Mahanadi River can be catalogued.

Acknowledgement The authors are thankful to the R&D Division, DoWR, RD & GR, Ministry of Jal Shakti, Government of India for providing financial assistance (No. 10/28/2022-R&D/, July 26, 2023) to sponsor the conference where this paper was presented.

References

Baliarsingh B, Kosygin L (2017) A new species of *Parapsilorhynchus* Hora, 1921 (Teleostei, Cyprinidae) from Mahanadi River basin of Odisha, India. Indian J Fish 4(4):44–49. https://doi.org/10.21077/ijf.2017.64.4.35503-06

Baliarsingh B, Kosygin L, Swain S (2017) *Parapsilorhynchus odishaensis*, a new cyprinid fish (Teleostei: Cyprinidae) from Odisha, India. Rec Zool Surv India 117(1):22–25. https://doi.org/10.26515/rzsi/v117/i1/2017/117278

Baliarsingh K, Swain SK, Oliver KED, Nandi D, Rath B (2020) Distribution and status of freshwater fish fauna and its habitat in the water bodies of Kendrapara district, Odisha, India. Biol Forum An Int J 12(2):44–50

Bharati PAS, Behera M, Naik K, Priyadarshini B, Raita SR, Kadraka LK, Prasad S, Kumar S (2019). Indigenous fishes of Odisha. J Biod Conserv 3(1): 211–219

Bhatt JP, Manish K (2023) A review on the threatened species of snow trout *Schizothorax richardsonii* gray, 1832 (Cypriniformes: Cyprinidae): from climate change and conservation perspectives. Indian J Fish 70(2):148–155. https://doi.org/10.21077/ijf.2023.70.2.131861-20

Bhatt JP, Manish K, Pandit MK (2012) Elevational gradients in fish diversity in the Himalaya: water discharge is the key driver of distribution patterns. PLoS One 7(9). https://doi.org/10.1371/journal.pone.0046237

Choubey K, Qureshi Y (2013) Study of ichthyofaunal biodiversity of Rajnandgaon town, CG, India. Int Res J Biol Sci 2(2):21–24

Dahanukar N, Raut R, Bhat A (2003) Distribution, endemism and threat status of freshwater fishes in the Western Ghats of India. J Biogeogr 31:123–136. https://doi.org/10.1046/j.0305-0270.2003.01016.x

Dash S, Payra A, Payra P, Palei H, Mishra RK, Mishra A (2018) Ichthyofaunal diversity in Ansupa lake, Cuttack, Odisha, India. e-Planet 16(2):156–164. https://doi.org/10.13140/RG.2.2.18239.76965

Díaz G, Górski K, Manosalva A, Toledo B, Habit E (2023) Fragmentation level drives local fish assemblage diversity patterns in fragmented river basins. Diversity 15(3):352. https://doi.org/10.3389/fenvs.2022.943586

Gopi KC, Kosygin L, Chandra K, Raha A (2018) Vertebrata: pisces. In: Fauna of Barnawapara wildlife sanctuary, Chhattisgarh. Zoological Survey of India, Kolkata

Hora SL (1940) On a col1ection of fish from the headwaters of the Mahanadi River, Raipur district, C.P. Rec Indian Mus 42(2):365–374

India-WRIS.2012, River Basin Atlas of India, RRSC-West, NRSC, ISRO, Jodhpur, India

Johnson J, Nanda P, Pant B, Mane S, Kolipakam V (2023) Redescription of an endemic mahseer, *Tor mahanadicus* (David, 1953) from Mahanadi river basin, India based on integrated morphological and molecular techniques. PLoS One 18(9). https://doi.org/10.1371/journal.pone.0291436

Kand N, Harit AK, Singh M, Bhhattacharyya S (2023) An assessment of freshwater fish diversity of Nuapada district, Odisha. J Surv Fish Sci 10(1S):6663–6671. https://doi.org/10.53555/sfs.v10i1S.2237

Khanani AB, Aggarwal RK (2018) The study on ichthyofaunal diversity in Chhattisgarh. J Emerg Technol Innov Res 5(10):604–611

Lall NA (2020) Fisheries of Mahanadi basin of Chhattisgarh. Int J Rev Res Soc Sci 8(1):11–12. https://doi.org/10.53550/EEC.2023.v29i04.031

Li J, Hang L, Sato T, Zou L, Jiang K, Yahara T, Kano Y (2012) Distribution pattern, threats and conservation of fish biodiversity in the East Tiaoxi, China. Environ Biol Fish 96:519–533. https://doi.org/10.1007/s10641-012-0036-z

Mishra K, Nayak PL (2014) A study of water pollution in two major rivers in Odisha-Mahanadi and Brahmani. Middle-East J Sci Res 22(12):1760–1770. https://doi.org/10.5829/idosi.mejsr.2014.22.12.21634

Moraga D, Vivancos A, Ruiz VH, Rojas O, Díaz G, Manosalva A, Vega P, Habit E (2022) A century of anthropogenic river alterations in a highly diverse river coastal basin: effects on fish assemblages. Front Environ Sci 10. https://doi.org/10.3389/fenvs.2022.943586

Nandini SB, Parida SP (2020) Studies on fish diversity of Mahanadi, nearby Zobra anicut, Cuttack, Odisha, India. J Emerg Technol Innov Res 7(4):881–885

Niyazi A, Swarnkar S, Keer NR, Sahu D, Singh J (2020) Ichthyofaunal diversity of Kharung reservoir in Bilaspur district, Chhattisgarh, India. J Exp Zool India 23(2):1317–1323

Oberdoff T, Guégan J-F, Hugueny B (1995) Global scale patterns of fish species richness in rivers. Ecography 18:345–352. https://doi.org/10.1111/j.1600-0587.1995.tb00137.x

Oberdorff T, Tedesco PA, Hugueny B, Leprieur F, Beauchard O, Brosse S, Dürr HH (2011) Global and regional patterns in riverine fish species richness: a review. Int J Ecol 2011(1). https://doi.org/10.1155/2011/967631

Panda J, Sahoo S (2020) Fish diversity and water quality of wetlands in Attabira block of Bargarh District, Odisha, India. Int J Ecol Environ Sci 2(2):66–67

Patel G, Chari MS, Kumar S, Bhakta D, Behera S, Verma NK, Chawan RR, Kumar V, Ahmad T (2016a) Fish fauna diversity of Mahanadi river in Raigarh district, Chhattisgarh. J Exp Zool India 19(1):1285–1289

Patel G, Chari MS, Kumar S, Bhakta D, Behera S (2016b) Status of ornamental fish diversity of Raigarh district, Chhattisgarh, India. Int J Sci Nat 7(3):575–578

Patel G, Chari MS, Kumar S, Bhakta D, Behera S, Verma NK, Chavan RR, Ahmad T (2016c) Ichthyofaunal diversity of three rivers of Raigarh district, Chhattisgarh. Natl J Life Sci 13(1):95–99

Pradhan B, Rajendran A, Khatua KK (2024) Understanding morphological dynamics of the Mahanadi river reach. J Geol Soc India 100(3):355–366. https://doi.org/10.17491/jgsi/2024/173842

Ratha KC (2019) Growing industrialization and its cumulative impacts: insights from the Mahanadi River basin. Soc Change 49(3):519–526. https://doi.org/10.1177/0049085719863904

Rejani C, Mandal S, Mayekar T, Bisht A, Singh S, Tyagi L (2023) Fish diversity and habitat ecology of Tel river, a tributary of Mahanadi river. J Inland Fish Soc India 54(1):03–15. https://doi.org/10.47780/jifsi.54.1.2022.132366

Sajina AM, Nair SM, Sudheesan D, Samanta S, Paul SK, Bhowmick S, Kumar V, Das BK (2024) Reconnoitre on ichthyofauna of Mahanadi river of India: shifting diversity down the river continuum and linking ecological traits with patterns in biodiversity. Environ Sci Pollut Res 31:5684–5698. https://doi.org/10.1007/s11356-023-31411-w

Singh TK (2014) Fish diversity in selected stretch of the river Mahanadi in Odisha and the livelihood of inhabiting fisher community. Int Res J Biol Sci 3:98–104

Singh RK, Das M (2018) Mahanadi: the great river. In: Singh D (ed) The Indian rivers. Springer Hydrogeology, Singapore, pp 309–318

Singh S, Omprakash C, M. and Vardia H. (2006) Diversity of fish fauna in catchment of Mahanadi river in Raipur district of Chhattisgarh. Environ Ecol 24(1):165–169

Singh TK, Saroj SK, Guru BC (2015) Biodiversity of ornamental fishes in the Hirakud reservoir and their socioeconomic benefits to fishers' communities of the region. Indian J Nat Sci 6(31):9074–9082

Singh TK, Swain SK, Guru BC (2020) Fish diversity of Mahanadi river (Odisha part), threats and conservation measures. Int J Life Sci 8(2):355–371

Spurgeon JJ, Pegg M, Parasiewicz P, Rogers J (2018) Diversity of river fishes influenced by habitat heterogeneity across hydrogeomorphic divisions. River Res Appl 34(7):797–806. https://doi.org/10.1002/rra.3306

Swain S (2014) Impact of climate variability over Mahanadi river basin. Int J Eng Res Technol 3:38–943

Tamboli RK, Jha YN (2012) Status of Catfish diversity of river Kelo and Mand in Raigarh district, CG, India. ISCA J Biol Sci 1(1):71–73

Tandi S, Tandi S (2018) Livelihood of fishing community and water pollution of Mahanadi river: A case study in Odisha. Asian J Multidimensio Res 7(9):107–109

Tyagi LK, Chandran R, Mandal S, Mayekar TS, Bisht AS, Singh SK (2020) People's perception on the status of fisheries resources and habitats in Mahanadi river basin. J Commun Mobiliz Sustain Dev 15(1):39–48

Tyagi LK, Chandran R, Mandal S, Mayekar TS, Bisht AS, Singh SK, Lal KK (2021) A checklist of fishes of Mahanadi river, India. J Indian Fish Assoc 48(1):1–11

Pathways of Social Inclusion for Soil and Water Conservation in the Mahanadi River Basin

7

Sarita Mishra, Ajay Kumar Mishra, and Jay Prakash Bhatt

Abstract

The Mahanadi River Basin is a critical lifeline for the eastern Indian states of Odisha and Chhattisgarh, supporting millions of livelihoods through agriculture, fishing, and other domestic uses. However, this basin faces increasing ecological and socioeconomic challenges due to rapid industrialization, urban expansion, and climate change. This chapter explores the integration of social inclusion into soil and water conservation efforts within the Mahanadi River Basin. By engaging marginalized groups such as tribal communities, women, and landless farmers in conservation initiatives, social inclusion fosters a sense of ownership and ensures that solutions are locally adapted and sustainable. The manuscript emphasizes that involving these communities in planning and implementing conservation practices significantly improves the outcomes and long-term success of such initiatives. Through a comprehensive review of current conservation challenges, successful case studies, and policy frameworks, this chapter proposes practical strategies for fostering inclusivity in environmental management, ultimately aiming to increase the ecological and socioeconomic resilience of the Mahanadi River Basin.

Keywords

Mahanadi River Basin · Social inclusion · Soil conservation · Water conservation · Climate change · Marginalized communities · Sustainable management · Community participation · Equity · Environmental resilience

S. Mishra (✉) · J. P. Bhatt
Faculty of Science, Sri Sri University, Cuttack, Odisha, India

A. K. Mishra
International Rice Research Institute, South Asia Regional Centre, Varanasi, Uttar Pradesh, India

J. P. Bhatt, A. K. Mishra (eds.), *Mahanadi River*,
https://doi.org/10.1007/978-981-96-1048-8_7

7.1 Introduction

The Mahanadi River Basin, which extends across the states of Chhattisgarh and Odisha in eastern India, is one of the most vital water systems in the region. Covering an area of approximately 141,600 square kilometers, the river stretches over 858 kilometers, supporting diverse ecosystems and sustaining millions of livelihoods. Historically, Mahanadi has been the lifeline for communities that are dependent on agriculture, fishing, and various domestic uses. Its waters nourish fertile lands, making the basin an agricultural hub that significantly contributes to the regional economy. Additionally, river ecosystems support rich biodiversity, including a variety of aquatic species, which are integral to local fishing communities (Jin et al. 2018; Ratha 2019; Pany et al. 2023; Sajina et al. 2024). However, over the past few decades, the Mahanadi River Basin has faced increasing ecological and environmental challenges. Rapid industrialization, urban expansion, and the overextraction of water resources for industrial and agricultural needs have led to a decline in river health. Issues such as reduced water flow, increasing pollution levels, and alterations in natural drainage patterns have resulted in the degradation of the basin's water quality and ecosystem balance. The situation is further exacerbated by the impacts of climate change, which has led to unpredictable weather patterns, frequent flooding, and water scarcity during dry seasons. These challenges pose a significant threat not only to the ecological balance of the basin but also to the socioeconomic well-being of communities that rely on rivers (Jin et al. 2018; Samuel et al. 2017; Samanta et al. 2020). Given the critical role of the Mahanadi River Basin, there is an urgent need for sustainable management and conservation strategies to ensure the long-term health and productivity of this water system. One of the key aspects of sustainable water and soil conservation is the integration of social inclusion in these initiatives. Social inclusion involves ensuring that all stakeholders, particularly marginalized and vulnerable groups, are actively engaged in conservation efforts. By involving local communities, including farmers, fisherfolks, tribal populations, and women, in the planning and implementation of conservation programs, there is a greater likelihood of success. Engaging these communities fosters a sense of ownership, enhances local knowledge, and ensures that conservation measures are tailored to the specific needs and conditions of the basin (Raut et al. 2023; Pahl-Wostl and Knieper 2014). The scope of social inclusion in soil and water conservation within the Mahanadi River Basin is vast and multifaceted. It involves not only technical solutions, such as sustainable agricultural practices, water management, and reforestation, but also policies that promote equitable access to resources, education, and capacity building. Research has shown that when conservation initiatives are inclusive, they are more effective and sustainable in the long term. For example, the involvement of local communities in watershed management can improve water use efficiency, prevent soil erosion, and increase groundwater recharge. Furthermore, empowering communities through education and capacity building can lead to innovative local solutions for tackling environmental challenges.

This study explores the importance of integrating social inclusion in soil and water conservation efforts within the Mahanadi River Basin. It analyzes the current ecological and socioeconomic challenges facing the basin, assesses the barriers to effective and inclusive conservation, and presents practical strategies that have proven successful in similar contexts. By highlighting the principles of social inclusion—such as equity, participation, recognition, and empowerment—the manuscript aims to provide a comprehensive framework for enhancing sustainable water and soil management practices in the region. Through case studies and examples from the Mahanadi River Basin and other similar ecological regions, this study emphasizes the need for a collaborative approach that brings together stakeholders from various sectors, including government agencies, nongovernmental organizations, and local communities. Ultimately, the goal is to promote the development of policies and practices that not only address the environmental concerns of the basin but also improve the livelihoods and resilience of the communities that depend on it.

7.1.1 Critical Role of the Mahanadi River Basin in Regional Ecology and Economics

The Mahanadi River Basin plays a crucial role in the regional ecology and economy of Chhattisgarh and Odisha (Fig. 7.1). The river spans approximately 858 kilometers and is vital for agriculture, fisheries, and domestic use, supporting millions of people who depend on its resources for their livelihoods. The basin is characterized by diverse ecological zones ranging from forested areas to agricultural lands,

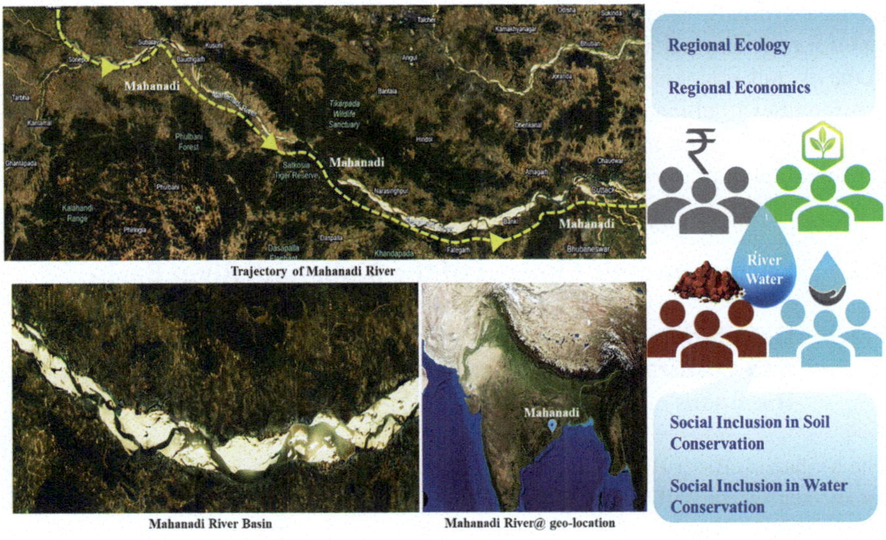

Fig. 7.1 Crucial importance of the Mahanadi River Basin for regional environmental and economic stability

supporting millions of people who depend on its resources for their livelihoods. However, the basin is now faced with significant ecological challenges due to rapid industrialization and the overallocation of water to industries (Ratha 2019). River flow and health are being impacted, leading to environmental and ecological consequences. Furthermore, the basin is vulnerable to climate change, with alterations in river flows and deteriorating water quality being key concerns (Rao 1995). These factors highlight the urgent need for sustainable management and conservation efforts in the Mahanadi River Basin.

7.1.2 Scope and Importance of Social Inclusion in Soil and Water Conservation Initiatives

The scope and importance of social inclusion in soil and water conservation initiatives are multifaceted. Sustainable soil management practices, as emphasized by Altobelli et al. (2020), are critical in addressing soil degradation, and their success can be significantly enhanced through social inclusion. Gidley et al. (2010) highlighted the importance of considering various areas and degrees of inclusion, such as socioeconomic status and community participation, in these initiatives. Raut et al. (2023) underscored the challenges of implementing social inclusion policies in community water resource management, suggesting the need for a nuanced approach. Bhattacharyya et al. (2016) also emphasized the role of watershed programs in addressing soil and water conservation, which can be strengthened through social inclusion strategies.

Soil and water conservation are critical for maintaining the health and sustainability of the Mahanadi River Basin. These practices help prevent soil erosion, increase groundwater recharge, maintain soil fertility, and ensure the availability of water for agricultural and domestic use. Effective conservation strategies contribute to the overall ecological balance, support biodiversity, and increase the resilience of communities to climate change impacts. Collectively, these studies underscore the importance of social inclusion in promoting effective and equitable soil and water conservation initiatives.

7.2 Socioeconomic Context of the Mahanadi River Basin

7.2.1 Detailed Overview of the Demographic, Economic, and Cultural Characteristics of the Communities Within the Basin

The Mahanadi River Basin, which spans Chhattisgarh and Odisha, is a vital resource for the region, providing water for irrigation and domestic and industrial use. It is home to more than 55 million people, with a diverse population comprising various ethnic groups, including tribal communities, farmers, and fisherfolk, supporting the livelihoods of local fishing communities. However, these communities often face

Fig. 7.2 Socioeconomic landscape of the Mahanadi River Basin

socioeconomic challenges, including poverty and vulnerability to climate change (Tripathy and Mondal 2019). River water quality is also a concern, with pollution and seasonal variability posing significant risks. The demographic composition of the basin includes a mix of urban and rural populations, with varying degrees of access to resources and services. These factors underscore the need for sustainable management and targeted policies to improve the well-being of the communities within the basin (Fig. 7.2).

7.2.2 The Dependence of These Communities on the Basin's Natural Resources for Livelihoods, Including Agriculture, Fishing, and Tourism

The Mahanadi River Basin is home to diverse communities with a strong dependence on its natural resources for livelihoods, including agriculture, fishing, and tourism. Agriculture is the primary livelihood for many communities, with rivers providing essential water for irrigation. Fishing is another significant activity that supports numerous households along the river's course. Tourism, although less developed, offers potential for economic growth, particularly in areas with rich cultural and natural heritage. However, these communities face significant challenges, including poverty and vulnerability to climate change impacts (Jin et al. 2018). The socioeconomic status of these communities is a key factor in shaping their relationship with the basin, with issues such as population growth, urbanization, and poverty influencing their access to and use of natural resources (Rai et al. 2012). Thus, there is a need for targeted policies and sustainable development initiatives to improve the socioeconomic conditions of these communities and ensure the long-term health of the basin (Behera et al. 2023).

7.3 Challenges to Inclusive Conservation

7.3.1 Barriers to Social Inclusion in Current Conservation Efforts

The barriers to social inclusion in conservation efforts are complex and multifaceted. Johri et al. (2021) and Paulson et al. (2012) both highlighted the unequal power relations between indigenous peoples and the Global North, as well as the limited resources for broad inclusion. Resource access disparities among communities pose significant challenges to inclusive conservation. Marginalized groups, including tribal communities and landless farmers, often face limited access to land, water, and other resources essential for their livelihoods. These disparities hinder their participation in and benefits from conservation initiatives. Smith et al. (2017) underscored the need for workforce diversity and the establishment of a marine diversity network to alleviate these barriers. Matulis and Moyer (2017) argue for the value of pluralism and the need for agonism in maintaining space for historically underrepresented perspectives.

Existing policies may not adequately address the needs of all community members, particularly marginalized groups. Inadequate policy frameworks and a lack of enforcement mechanisms can exacerbate inequalities and undermine conservation efforts. Thus, ensuring that policies are inclusive and effectively implemented is crucial for successful conservation.

These studies collectively underscore the need to address disparities in resource access, policy limitations, and lack of community engagement to achieve inclusive conservation.

7.3.2 The Impact of These Challenges on the Effectiveness and Sustainability of Conservation Initiatives

A range of challenges impact the effectiveness and sustainability of conservation initiatives. Hoffmann (2021) and Negacz et al. (2022) highlight the need for improved governance and management of protected areas, particularly in the face of human land use, climate change, and invasive species. They also stress the potential of international cooperative initiatives to address these challenges. Dawson et al. (2021) emphasize the importance of indigenous peoples and local communities in conservation efforts, suggesting that their involvement can lead to more effective and equitable outcomes. Redpath et al. (2013) underscored the need for effective conflict management to minimize negative impacts on biodiversity and human well-being. Overall, these studies underscore the need for increased support and resources, as well as improved governance and management, to enhance the effectiveness and sustainability of conservation initiatives.

7.4 Principles of Social Inclusion in Conservation

7.4.1 Social Inclusion in the Context of Environmental Management

Social inclusion in environmental management encompasses a range of social issues, including multiculturalism, gender inequality, low income, and racialization. It is particularly important in the context of sustainable urban development, where it is often an autonomous component rather than mainstream (Mirzoev et al. 2021). Strategies for the sustainable management of solid urban waste should prioritize social inclusion, particularly in the context of independent waste gatherers (Gouveia 2012). However, the concept of social inclusion is context-dependent, which can vary depending on the country, region, and locality (Silver 2015). Therefore, social inclusion in environmental management involves ensuring that all community members, regardless of their social, economic, or cultural background, have equal opportunities to participate in and benefit from conservation efforts. It encompasses principles of equity, participation, recognition, and empowerment.

7.4.2 Principles Necessary to Foster Inclusivity, Such as Equity, Participation, Recognition, and Empowerment

Inclusive practices, encompassing equity, participation, recognition, and empowerment, are essential for a fair and just society (Fig. 7.3). In education, these practices can enhance academic, social, and emotional development, but they often require overcoming challenges such as limited resources and attitudinal barriers (Pradhan and Naik 2024). The principle of inclusion, as the opposite of exclusion and alienation, necessitates the removal of barriers and the facilitation of participation in all areas of life (Jones 2010). In the field of interaction design and children, researchers are exploring how to promote equity and inclusivity, particularly in relation to diverse children and families (Sobel et al. 2017).

7.5 Strategies for Enhancing Social Inclusion

7.5.1 Detailed Practical Strategies and Interventions Aimed at Integrating Marginalized Groups, Enhancing Stakeholder Collaboration, and Ensuring Equitable Benefit Sharing

A range of strategies have been proposed to enhance social inclusion, particularly for marginalized groups, such as improving accessibility to public services, developing inclusive infrastructure, and empowering the community through training programs. In the context of sports, strategies such as developing unified sports and providing training to coaches have been identified as effective in fostering inclusion

Fig. 7.3 Core principles of social inclusion in conservation

for people with intellectual disabilities (Grandisson et al. 2019). Moreover, the need for systemic change and innovative approaches to social inclusion at the institutional and community levels should be emphasized. Rimmerman (2013) highlights the importance of national and international strategies, including social protection, disability rights instruments, and media representation, in promoting the social inclusion of people with disabilities. These strategies collectively underscore the importance of collaboration and equitable benefit sharing in enhancing social inclusion.

7.5.2 Highlights of Successful Case Studies from Similar Ecological Regions Where Inclusive Practices Have Led to Improved Conservation Outcomes

Inclusive conservation practices have been shown to significantly improve conservation outcomes in various ecological regions. Cebrián-Piqueras et al. (2023) identified key leverage points and levers for inclusive conservation, including improvements in the sociocultural context, social cohesion, and the empowerment of local communities. Huber et al. (2023) further emphasize the importance of genuine devolution of power, diverse actor involvement, long-term external support, and devolution of rights in achieving successful outcomes. These findings are supported by Foggin (2018), who also highlights the critical role of community engagement and participation in the success of conservation ventures in the Tibetan Plateau region. Chaplin-Kramer et al. (2023) underscored the importance of recognizing and respecting local values and knowledge in terms of achieving positive impacts for both nature and people. These studies collectively demonstrate the transformative potential of inclusive conservation practices in improving conservation outcomes.

7.6 Community-Based Conservation Models

7.6.1 Models of Conservation That Prioritize Community Leadership and Participation, Such as Comanagement and Community-Owned Resources

Community-based conservation models, such as comanagement and community-owned resources, have gained prominence in recent years. These models prioritize community leadership and participation, aiming to reconcile human development with ecosystem sustainability (Timmer 2004). In China, community participation in conservation has been classified into three modes: community participatory management, community comanagement, and community dominant management, with the success of these models hinging on stable land tenure and institutional changes (He et al. 2020). Similarly, Qianxiang et al. (2013) highlighted the importance of community-based integrated watershed management approaches in enhancing ecosystem service values. These models prioritize community leadership and participation, thus emphasizing the importance of collaboration, local empowerment, and sustainable resource management.

7.6.2 Examples and Lessons Learned from the Implementation of These Models in Different Regions

Various regions have successfully implemented community-based conservation models, offering valuable lessons. In sub-Saharan Africa, community-based forest management (CBFM) has been promoted as a means of improving forest conditions and rural livelihoods, although challenges such as weak tenure rights and local elite capture remain (Duguma et al. 2018). In the Kavango–Zambezi Transfrontier Conservation Area (KAZA TFCA), the KAZA Elephant Conservation and Management Strategy highlights the importance of stakeholder involvement and the sustainable use of wildlife resources. In the Pacific Island region, community-based conservation initiatives have successfully integrated local knowledge and customs into conservation planning, resulting in improved biodiversity outcomes and strengthened community resilience. These examples underscore the importance of inclusive participation, equitable benefit sharing, and adaptive management in the successful implementation of community-based conservation models.

7.7 Policy and Institutional Frameworks

7.7.1 Existing Policy Frameworks Governing Soil and Water Conservation in the Mahanadi River Basin

The existing policy frameworks governing soil and water conservation in the Mahanadi River Basin are shaped by various national and regional initiatives. The Integrated Watershed Management Programme (IWMP) is a key policy initiative aimed at promoting sustainable water and soil management in India. The National Water Policy (2012) and the National Environment Policy (2006) also provide a comprehensive framework for water conservation and management, emphasizing the importance of integrated watershed development and community participation. Additionally, the Odisha State Water Policy (2007) and the Chhattisgarh State Water Policy (2012) outline specific strategies for sustainable water management in the Mahanadi River Basin, including measures for soil conservation and groundwater recharge. These policy frameworks collectively aim to address the challenges of water scarcity, soil erosion, and environmental degradation in the Mahanadi River Basin through sustainable and inclusive approaches (Table 7.1).

Table 7.1 Policy frameworks for soil and water conservation in the Mahanadi River Basin

Policy framework	Description	Key features
Integrated Watershed Management Programme (IWMP)	A key initiative aimed at promoting sustainable water and soil management in India.	Focuses on integrated watershed development, community participation, and sustainable resource management.
National Water Policy (2012)	Provides a comprehensive framework for water conservation and management in India.	Emphasizes integrated watershed development, community involvement, and sustainable water management practices.
National Environment Policy (2006)	Outlines strategies for environmental protection, including water management.	Highlights the importance of ecosystem conservation, sustainable development, and community participation.
Odisha State Water Policy (2007)	Specifies strategies for sustainable water management in Odisha, particularly in the Mahanadi River basin.	Includes measures for soil conservation, groundwater recharge, and community engagement.
Chhattisgarh State Water Policy (2012)	Focuses on sustainable water management practices specific to Chhattisgarh, including the Mahanadi River basin.	Emphasizes integrated approaches, community participation, and strategies for soil and water conservation.

7.7.2 Gaps and Areas for Improvement in These Policies to Enhance Social Inclusion

Despite the comprehensive policy frameworks governing soil and water conservation in the Mahanadi River Basin, there are several gaps and areas for improvement to enhance social inclusion. The National Environment Policy (2006) and the Odisha State Water Policy (2007) emphasize the importance of community participation, but their implementation often falls short due to limited resources and inadequate institutional support. Similarly, the Chhattisgarh State Water Policy (2012) outlines strategies for sustainable water management but lacks specific provisions for addressing the needs of the marginalized communities. To enhance social inclusion, these policies need to incorporate more explicit measures for equitable benefit sharing, capacity building, and the empowerment of marginalized groups. Strengthening institutional frameworks, improving resource allocation, and fostering multistakeholder collaboration are also critical for ensuring the effective implementation of these policies (Table 7.2).

7.8 Technological and Innovative Approaches to Support Inclusion

7.8.1 Role of Technology in Inclusive Conservation Strategies

The use of technology, particularly geographic information systems (GISs), remote sensing, and mobile technology, has significantly enhanced inclusive conservation

Table 7.2 Gaps and recommendations for enhancing social inclusion in soil and water conservation policies in the Mahanadi River Basin

Policy framework	Gaps/areas for improvement	Recommendations
National Environment Policy (2006)	Limited implementation of community participation due to insufficient resources and inadequate institutional support.	Enhance resource allocation and strengthen institutional frameworks for better community engagement.
Odisha State Water Policy (2007)	Emphasis on community participation is not effectively realized; lacks measures to address the needs of marginalized communities.	Incorporate explicit measures for equitable benefit sharing and capacity building for marginalized groups.
Chhattisgarh State Water Policy (2012)	Strategies for sustainable water management are insufficiently focused on the inclusion of marginalized communities.	Develop specific provisions targeting the empowerment and inclusion of marginalized groups in water management initiatives.
Overall recommendations	Need for strengthening institutional frameworks and fostering multistakeholder collaboration.	Promote collaborative approaches that involve government, community, and nongovernmental stakeholders in policy implementation.

strategies. Beckline et al. (2017) highlighted the role of these technologies in monitoring land cover changes, assessing threats, and evaluating forest management patterns. These tools provide critical data for making informed decisions and engaging communities in conservation efforts. Menon and Bawa (1997) further emphasized their application in biodiversity conservation, including deforestation rate estimation and conservation planning. However, Adams (2019) raised concerns regarding the potential negative implications of these technologies, such as the demarcation and control of space and the rise of coercive conservation strategies. Despite these concerns, the overall impact of technology on inclusive conservation strategies is positive, as it enables more effective monitoring and management of natural resources.

7.8.2 Innovative Financing Mechanisms to Incentivize Community Participation

Innovative financing mechanisms, such as payment for ecosystem services (PES), can incentivize community participation in environmental protection (Pojani 2021; Aryal et al. 2019). PES compensates individuals or communities for actions that enhance ecosystem services and has been found to be effective in watershed management in Albania and Nepal (Pojani 2021; Aryal 2019). However, the success of PES depends on its design, which should consider the environmental, socioeconomic, and political context. In the context of the Hindu Kush Himalaya, an incentive-based mechanism, rather than a purely market-based mechanism, is recommended to ensure long-term benefits for local communities (Bhatta et al. 2014).

The following are some of the financing mechanisms:

7.8.2.1 Payment for Ecosystem Services (PES)

Payment is a market-based approach that incentivizes landowners and communities to engage in conservation activities by providing financial compensation for the ecosystem services they provide:

- *Concept*: PES schemes compensate individuals or communities for managing their land or water resources in ways that generate ecosystem services, such as carbon sequestration, water purification, and biodiversity conservation.
- *Implementation*: In the Mahanadi River Basin, PES could be used to encourage sustainable agricultural practices, reforestation, and wetland conservation. Farmers and landowners receive payments for adopting practices that enhance soil health and water quality.
- *Benefits*: PES provides a direct financial incentive for conservation, making it economically viable for communities to protect and restore ecosystems. It also

promotes long-term sustainability by linking economic benefits to environmental stewardship.

7.8.2.2 Community-Based Microfinance

Community-based microfinance involves providing small loans and financial services to low-income individuals and groups to support conservation-related activities:

- *Microfinance Institutions (MFIs)*: MFIs can offer loans to farmers and community groups for investments in sustainable agricultural practices, water-saving technologies, and other conservation initiatives.
- *Community Savings Groups*: These groups enable members to pool their savings and provide loans to each other for conservation projects. This fostered a sense of ownership and collective responsibility for local conservation efforts.
- *Impact*: Access to microfinance empowers communities to implement conservation practices that they might not otherwise afford. It also enhances their financial resilience and capacity to invest in sustainable livelihoods.

7.8.2.3 Crowdfunding

Crowdfunding is a method of raising funds through small contributions from a large number of people, typically via online platforms:

- *Platforms and Campaigns*: Conservation organizations and community groups can use crowdfunding platforms to raise money for specific conservation projects. Campaigns can highlight the environmental and social benefits of projects, attracting support from a broad audience.
- *Community Engagement*: Crowdfunding campaigns can engage and mobilize local and global communities, raising awareness and fostering a sense of collective action. Supporters can contribute financially and help spread the word about the campaign.
- *Transparency and Accountability*: Crowdfunding platforms often provide tools for tracking progress and reporting on the use of funds, ensuring transparency and accountability. These factors build trust among contributors and encourage ongoing support for conservation initiatives (Table 7.3).

7.9 Monitoring, Evaluation, and Adaptive Management

7.9.1 Methods for Monitoring the Impacts of Socially Exclusive Practices on Conservation Outcomes

A range of methods have been proposed for monitoring the impacts of socially inclusive practices on conservation outcomes. Kaplan and Bennett (2018) suggested the use of adaptive social impact management, which involves a cyclical process of monitoring and managing social impacts. This approach can be complemented by

Table 7.3 Technological and innovative approaches to enhance inclusion in conservation strategies

Approach	Description	Implementation examples	Benefits
Role of technology in inclusive conservation strategies	Technology enhances conservation efforts through tools like GIS, remote sensing, and mobile technology. These tools help monitor land-cover changes and assess environmental threats	Monitoring land-cover changes, evaluating forest management patterns	Provides critical data for informed decision-making and community engagement
Innovative financing mechanisms	Financing mechanisms like payment for ecosystem services (PES) incentivize community participation in environmental protection by compensating for actions that enhance ecosystem services	PES in watershed management in Albania and Nepal	Encourages community participation and aligns economic benefits with environmental stewardship
Payment for ecosystem services (PES)	A market-based approach compensating landowners/communities for ecosystem services such as carbon sequestration and biodiversity conservation	PES in Mahanadi River basin to promote sustainable practices	Direct financial incentive for conservation, promoting long-term sustainability
Community-based microfinance	Provides small loans and financial services to low-income individuals and groups to support conservation-related activities	MFIs offering loans for sustainable agriculture and conservation projects	Empowers communities financially, enhancing resilience and enabling investments in sustainable practices
Crowdfunding	Raises funds through small contributions from a large audience, typically via online platforms, for specific conservation projects	Conservation campaigns using online platforms for fundraising	Engages local and global communities, ensuring transparency and building trust among contributors

the most significant change method, as proposed by Davies (2005), which standardizes the collection and analysis of stories of change. Crandall et al. (2018) outlined a variety of social research methods, such as surveys and participant observation, that can be used to inform conservation efforts. Nyaupane et al. (2020) emphasized the need for adaptive and politically explicit governance spaces in protected area management, highlighting the importance of community engagement and the potential for unintended outcomes.

7.9.2 Importance of Adaptive Management in Responding to Socioeconomic and Environmental Changes

Adaptive management is crucial in responding to socioeconomic and environmental changes, as it allows for the balancing of social priorities and the improvement of conservation targeting (Sims et al. 2014). However, its practical application has been limited, and it should be used only after disputing parties have agreed on a set of questions to be addressed. The importance of adaptive management is further highlighted in the context of river management in Kruger National Park, where it has led to a new stewardship model based on stakeholder involvement and learning-focused management (Pollard et al. 1970). Despite their potential, socioecological barriers can hinder the implementation of adaptive management, as seen in the case of Lake Fúquene in Colombia (Andrade et al. 2012).

7.10 Conclusion

7.10.1 Key Findings and Insights

The key findings and insights from the analysis of social inclusion in soil and water conservation in the Mahanadi River Basin underscore the critical role of inclusive conservation practices in achieving sustainable outcomes. Basin communities are highly dependent on their natural resources for livelihoods, and their effective participation in conservation initiatives is deemed essential. However, barriers to social inclusion, such as limited access to resources, policy gaps, and inadequate institutional support, hinder the effectiveness and sustainability of these efforts. Successful models of community-based conservation, such as comanagement and community-owned resources, highlight the importance of local leadership, equitable benefit sharing, and adaptive management. While comprehensive, policy frameworks require further refinement to address the needs of marginalized communities and enhance social inclusion. Overall, fostering inclusivity through equity, participation, recognition, and empowerment is vital for promoting sustainable soil and water conservation in the Mahanadi River Basin.

7.10.2 Recommended Steps for Implementing More Inclusive Conservation Practices in the Mahanadi River Basin

To implement more inclusive conservation practices in the Mahanadi River Basin, the following steps are recommended:

Strengthening community engagement: Enhance community participation in conservation initiatives through capacity-building programs, awareness campaigns, and inclusive decision-making processes.

Promote equitable benefit sharing: Ensure that the benefits of conservation efforts are equitably distributed among all community members, particularly marginalized groups.

Enhance policy frameworks: Refine existing policies to incorporate explicit measures for social inclusion, such as provisions for equitable access to resources, capacity building, and the empowerment of marginalized communities.

Foster multistakeholder collaboration: Promote the collaboration among government agencies, nongovernmental organizations, local communities, and other stakeholders to ensure a coordinated and inclusive approach to conservation.

Allocate adequate resources: Ensure sufficient funding and resources for the implementation of inclusive conservation practices with a focus on capacity building, infrastructure development, and institutional support.

Monitor and evaluate progress: Establish mechanisms for continuous monitoring and evaluation to ensure the effectiveness and sustainability of socially inclusive soil and water conservation initiatives in the Mahanadi River Basin.

References

Adams WM (2019) Geographies of conservation II: technology, surveillance and conservation by algorithm. Prog Hum Geogr 43:337–350

Altobelli F, Vargas R, Corti G, Dazzi C, Montanarella L, Monteleone A, Caon L, Piazza MG, Calzolari C, Munafò M, Benedetti A (2020) Improving soil and water conservation and ecosystem services by sustainable soil management practices: from a global to an Italian soil partnership. Ital J Agron 15(4):293–298

Andrade G, Franco L, Delgado J (2012) Socioecological barriers to adaptive management of Lake Fuquene, Colombia. Int J Des Nat Ecodyn 7:251–260

Aryal, K., Bhatta, L. D., Thapa, P. S., Ranabhat, S., Neupane, N., Joshi, J., Shrestha, K., & Shrestha, A. B. (2019). Payment for ecosystem services: Could it be a sustainable financing mechanism for watershed services in Nepal? Green Finance, 1(3), 221–236. https://doi.org/10.3934/GF.2019.3.221

Beckline M, Yujun S, Yvette B, Joh AB, Mor-Achankap B, Saeed S, Richard T, Wose J, Paul C (2017) Perspectives of remote sensing and GIS applications in tropical forest management. Am J Agric For 5:33

Behera R, Sharma N, Mohanty S, Sahoo S, Dash SS, Anand A, Pradhan M, Behera C, Adhikar B (2023) A comprehensive study of Kujimahal Village in Odisha for the purpose of socioeconomic and sustainable development. Ecol Environ Conserv 29:1177–1183

Bhatta LD, van Oort BE, Rucevska I, Baral H (2014) Payment for ecosystem services: possible instrument for managing ecosystem services in Nepal. Int J Biodivers Sci Ecosyst Serv Manage 10:289–299

Bhattacharyya R, Ghosh BN, Dogra P, Mishra PK, Santra P, Kumar S, Fullen MA, Mandal UK, Anil KS, Lalitha M et al (2016) Soil conservation issues in India. Sustainability 8(6):565

Cebrián-Piqueras MA, Palomo I, Lo V, López-Rodríguez MD, Filyushkina A, Fischborn M, Raymond C, Plieninger T (2023) Leverage points and levers of inclusive conservation in protected areas. Ecol Soc 28:18

Chaplin-Kramer R, Neugarten RA, González-Jiménez D, Ahmadia GN, Baird TD, Crane N, Delgoulet E, Eyster HN, Kurashima N, Llopis JC, Millington A, Pawłowska-Mainville A, Rulmal J, Saunders F, Shrestha S, Vaughan MB, Winter KB, Wongbusarakum S, Pascual U

(2023) Transformation for inclusive conservation: evidence on values, decisions, and impacts in protected areas. Curr Opin Environ Sustain 64:101347

Chhattisgarh State Water Resources Development Policy (2012). https://www.ielrc.org/content/e1233.pdf

Crandall SG, Ohayon JL, de Wit LA, Hammond JE, Melanson KL, Moritsch MM et al (2018) Best practices: social research methods to inform biological conservation. Aust J Environ Manag 25(1):6–23

Davies R.(2005) The 'Most significant change' (MSC) technique: a guide to its use

Dawson NM, Coolsaet B, Sterling EJ, Loveridge R, Gross-Camp ND, Wongbusarakum S, Sangha KK, Scherl LM, Phan HP, Zafra-Calvo N, Lavey WG, Byakagaba P, Idrobo CJ, Chenet A, Bennett NJ, Mansourian S, Rosado-May FJ. (2021) The role of Indigenous peoples and local communities in effective and equitable conservation. Ecology and Society, 26(3), Article 19. https://doi.org/10.5751/ES-12625-260319

Duguma L, Atela J, Ayana A, Alemagi D, Mpanda M, Nyago M, Minang P, Nzyoka J, Foundjem Tita D, Ntamag-Ndjebet C (2018) Community forestry frameworks in sub-Saharan Africa and the impact on sustainable development. Ecol Soc 23:21

Foggin JM (2018) Environmental conservation in the Tibetan plateau region: lessons for China's belt and road initiative in the mountains of Central Asia. Land 7:52

Gidley J, Hampson G, Wheeler L, Bereded-Samuel E (2010) Social inclusion: context, theory and practice. Aust J Univ Commun Engage 5(1):6–36

Gouveia N (2012) Solid urban waste: socioenvironmental impacts and prospects for sustainable management with social inclusion. Ciencia & saude coletiva 17(6):1503–1510

Grandisson M, Marcotte J, Niquette B, Milot É (2019) Strategies to Foster inclusion through sports: a scoping review. Early online

He S, Yang L, Min Q. (2020) Community participation in nature conservation: The Chinese experience and its implication to national park management. Sustainability, 12(11), 4760. https://doi.org/10.3390/su12114760

Hoffmann S (2021) Challenges and opportunities of area-based conservation in reaching biodiversity and sustainability goals. Biodivers Conserv 31:325–352

Huber JM, Newig J, Loos J (2023) Participation in protected area governance: a systematic case survey of the evidence on ecological and social outcomes. J Environ Manag 336:117593

Jin L, Whitehead PG, Rodda H, Macadam I, Sarkar S (2018) Simulating climate change and socio-economic change impacts on flows and water quality in the Mahanadi River system, India. Sci Total Environ 637–638:907–917

Johri S, Carnevale M, Porter L, Zivian AM, Kourantidou M, Meyer EL, Seevers J, Skubel RA (2021) Pathways to justice, equity, diversity, and inclusion in marine science and conservation. Front Mar Sci

Jones M. (2010) Inclusion, social inclusion and participation. In M. H. Rioux, L. A. Basser, & M. Jones (Eds.), Critical Perspectives on Human Rights and Disability Law (pp. 57–82). Brill | Nijhoff. https://doi.org/10.1163/ej.9789004189508.i-552.24

Kaplan-Hallam M, Bennett NJ (2018) Adaptive social impact management for conservation and environmental management. Conserv Biol 32(2):304–314

Matulis BS, Moyer JR (2017) Beyond inclusive conservation: the value of pluralism, the need for agonism, and the case for social instrumentalism. Conserv Lett 10:279

Menon S, Bawa KS (1997) Applications of geographic information systems, remote-sensing, and a landscape ecology approach to biodiversity conservation in the Western Ghats. Curr Sci 73:134–145

Mirzoev T, Tull KI, Winn N, Mir G, King N, Wright JM, Gong YY (2021) Systematic review of the role of social inclusion within sustainable urban developments. Int J Sustain Dev World Ecol 29:3–17

National Environment Policy (2006). https://ibkp.dbtindia.gov.in/DBT_Content_Test/CMS/Guidelines/20190411103521431_National%20Environment%20Policy,%202006.pdf

National Water Policy (2012). https://nwm.gov.in/sites/default/files/national%20water%20policy%202012_0.pdf

Negacz K, Petersson MT, Widerberg O, Kok MTJ, Pattberg PH. (2022). The potential of international cooperative initiatives to address key challenges of protected areas. Environmental Science & Policy, 136, 620–631. https://doi.org/10.1016/j.envsci.2022.07.026
Nyaupane GP, Poudel S, York AM (2020). Governance of protected areas: An institutional analysis of conservation, community, and tourism outcomes. Journal of Sustainable Tourism, 28(11), 1804–1824. https://doi.org/10.1080/09669582.2020.1762627
Pahl-Wostl C, Knieper C (2014) The capacity of water governance to deal with the climate change adaptation challenge: using fuzzy set qualitative comparative analysis to distinguish between polycentric, fragmented and centralized regimes. Glob Environ Chang 29:139–154
Pany R, Rath A, Swain PC (2023) Water quality assessment for river Mahanadi of Odisha, India using statistical techniques and artificial neural networks. J Clean Prod 417:137713
Paulson N, Laudati AA, Doolittle AA, Welsh-Devine M, Peña PR (2012) Indigenous peoples' participation in global conservation: looking beyond headdresses and face paint. Environ Values 21:255–276
Pojani E. (2021) Innovative financing mechanism for environmental protection: Applying PES in a watershed area in Albania. Interdisciplinary Journal of Research and Development, 8(1), 1–10
Pollard S, Toit DD, Biggs HC (1970) River management under transformation: the emergence of strategic adaptive management of river systems in the Kruger National Park. Koedoe 53:01–14
Pradhan DK, Naik MS (2024) Inclusive education: a foundation for equality and empowerment at the elementary stage. Int J Multidiscip Res Arts Sci Technol
Qianxiang L, Igbokwe K, Jiayong L (2013) Community-based integrated watershed management. Chin J Popul Resour Environ 3:60–64
Rai RK, Upadhyay A, Ojha CSP, Singh VP (2012) Socioeconomic status. In: The Yamuna River Basin, vol 66. Springer, Dordrecht, pp 141–164
Rao PG (1995) Effect of climate change on streamflows in the Mahanadi River Basin, India. Hydrol Sci J 40(4–5):503–517
Ratha KC (2019) Growing industrialization and its cumulative impacts: insights from the Mahanadi River Basin. Social Change 49(3):519–530
Raut M, Varady RG, Rajouria A (2023) Gender and social inclusion in community water resource management: lessons from two districts in the Himalayan foothills and the Terai in Nepal. Water Int 48(4):547–566
Redpath SM, Young JC, Evely A, Adams WM, Sutherland WJ, Whitehouse A, Amar A, Lambert RA, Linnell JD, Watt AD, Gutiérrez RJ (2013) Understanding and managing conservation conflicts. Trends Ecol Evol 28(2):100–109
Review of Odisha State Water Policy (2007) In line with National Water Policy, 2012 in context of climate change. https://cwp-india.org/wp-content/uploads/2018/03/Report-on-Odisha-SWP-with-regard-to-National-Water-Policy-2012.pdf
Rimmerman A. (2013) Social inclusion of people with disabilities: National and international perspectives. Cambridge University Press. https://doi.org/10.1017/CBO9781139035668
Sajina AM, Nair SM, Sudheesan D, Samanta S, Paul SK, Bhowmick S et al (2024) Reconnoitre on ichthyofauna of Mahanadi River of India: shifting diversity down the river continuum and linking ecological traits with patterns in biodiversity. Environ Sci Pollut Res 31(4):5684–5698
Samanta S, Kumar V, Nag SK, Raman RK, Saha K, Bandyopadhyay S et al (2020) Metal contaminations in sediment and associated ecological risk assessment of river Mahanadi, India. Environ Monit Assess 192:1–17
Samuel A, Joy KJ, & Bhagat S. (2017) Integrated water management of the Mahanadi Basin. In Water Conflicts in Odisha: A Compendium of Case Studies (pp. 123–145). Forum for Policy Dialogue on Water Conflicts in India.
Silver H (2015) The contexts of social inclusion. AARN: Class & Status (Sub-Topic)
Sims KRE, Alix-Garcia JM, Shapiro-Garza E, Fine LR, Radeloff VC, Aronson G, Castillo S, Ramirez-Reyes C, & Yañez-Pagans P. (2014). Improving environmental and social targeting through adaptive management in Mexico's payments for hydrological services program. Conservation Biology, 28(5), 1151–1159. https://doi.org/10.1111/cobi.12244

Smith, N. S., Côté, I. M., Martinez-Estevez, L., Hind-Ozan, E. J., Quiros, Á. L., Johnson, N., Green, S. J., Cornick, L. A., Shiffman, D. S., Malpica-Cruz, L., Gleason Besch, A., & Shiel-Rolle, N. (2017) Diversity and inclusion in conservation: A proposal for a marine diversity network. Frontiers in Marine Science, 4, Article 234. https://doi.org/10.3389/fmars.2017.00234

Sobel K, Kientz JA, Clegg TL, González C, Yip JC (2017) Equity & Inclusivity at IDC. In Proceedings of the 2017 conference on interaction design and children

Timmer VJ (2004) Community-based conservation and leadership: Frameworks for analysis. In Conservation Leadership: Managing for Change (pp. 45–67). Island Press.

Tripathy B, Mondal T (2019) Socioeconomic challenges faced by basin's people in India. Think India 22:296–304

Nature-Based Solutions for Mahanadi River Basin Protection and Conservation

8

Ajay Kumar Mishra, Sarita Mishra, Manas Ranjan Sahoo, and Sheetal Sharma

Abstract

The Mahanadi River Basin, a critical ecosystem in East Central India, supports millions of people through agriculture, fisheries, and freshwater resources. However, this basin faces several environmental challenges, including pollution, soil erosion, deforestation, and the adverse impacts of climate change. Nature-based solutions (NbS) offer a sustainable approach to address these issues by leveraging natural processes to restore ecosystems, mitigate floods, enhance water quality, and promote biodiversity. This chapter explores the potential of NbS for the conservation of the Mahanadi River Basin. Specific strategies such as riparian buffer zones, wetland restoration, sustainable agriculture, and community-led watershed management are discussed. By integrating these solutions into development planning, NbS can provide long-term environmental and socioeconomic benefits, including improved soil health, enhanced biodiversity, and climate resilience. This chapter emphasizes the importance of stakeholder engagement, community participation, and policy support in successfully implementing NbS in the Mahanadi River Basin to ensure its ecological and economic sustainability.

Keywords

Mahanadi River Basin · Nature-based solutions · Environmental conservation · Sustainable agriculture · Climate resilience

A. K. Mishra (✉) · M. R. Sahoo · S. Sharma
International Rice Research Institute (IRRI) South Asia Regional Centre, Varanasi, Uttar Pradesh, India
e-mail: a.k.mishra@irri.org

S. Mishra
Faculty of Science, Sri Sri University, Cuttack, Odisha, India

J. P. Bhatt, A. K. Mishra (eds.), *Mahanadi River*,
https://doi.org/10.1007/978-981-96-1048-8_8

8.1 Introduction

The Mahanadi River Basin, located in East Central India, is not only an ecological treasure but also a vital resource for millions of people who are dependent on it for agriculture, fishing, and other livelihoods. Spanning over 141,600 square kilometers, it is home to diverse flora and fauna, with several endemic species contributing to the region's rich biodiversity. The river provides irrigation to vast agricultural lands and supports inland fisheries, making it a key player in the regional economy. However, in recent decades, the basin has faced increasing environmental challenges, including soil erosion, water pollution, and the intensifying impacts of climate change. Industrial waste, agricultural runoff, and urban effluents severely degrade water quality, while deforestation and unregulated land use have exacerbated soil erosion and increased the basin's vulnerability to floods (Bohorquez et al. 2023; Debele et al. 2023).

To combat these growing threats, nature-based solutions (NbS) have emerged as promising approaches that utilize natural processes and ecosystems to address environmental and socioeconomic challenges. NbS strategies harness the power of nature to restore degraded landscapes, regulate water flow, and increase biodiversity, providing sustainable alternatives to traditional engineering solutions. NbS, by definition, integrates natural systems into development planning, offering multiple benefits, such as carbon sequestration, flood regulation, and enhanced water quality (FAO 2021). In the context of the Mahanadi River Basin, NbS could play a pivotal role in reversing environmental degradation while supporting the livelihoods of local communities (Scheme 8.1).

8.2 Nature-Based Solutions: A Global Perspective

Nature-based solutions (NbS) have gained considerable attention globally as effective strategies for addressing hydrometeorological risks such as floods, droughts, and landslides. The increasing frequency of extreme weather events due to climate change has highlighted the importance of sustainable and adaptive approaches that can offer long-term resilience. NbS represents an umbrella concept encompassing various ecosystem-based approaches, such as ecosystem-based adaptation (EbA), natural climate solutions (NCSs), and green infrastructure, which leverage the regenerative capacity of nature to provide societal benefits (Debele et al. 2023).

One of the primary reasons NbS has been embraced is its ability to provide cost-effective, sustainable, and flexible solutions compared with traditional gray infrastructure approaches. For example, wetlands and riparian zones can act as natural buffers against floods, whereas reforested areas can help reduce the intensity of landslides and soil erosion. Moreover, NbS is often seen as a holistic approach that provides cobenefits beyond disaster risk reduction, including biodiversity conservation, improved water quality, and enhanced carbon storage, all of which are critical in the fight against climate change (Keesstra et al. 2018).

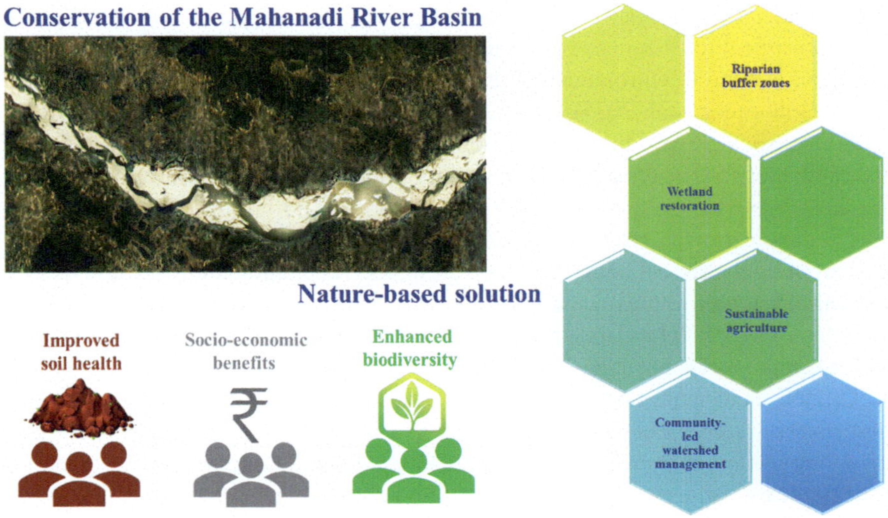

Scheme 8.1 Schematic overview of nature-based solutions for the Mahanadi River

Recent studies have shown that NbS can play a significant role in improving soil health and reducing landscape connectivity, which are deemed crucial in preventing erosion and retaining water during extreme weather events. By restoring the functionality of ecosystems, NbS promotes resilience at both the landscape and community levels, ensuring that natural resources continue to support human well-being even in the face of climate change. In the Hindu Kush Himalayas, for example, NbS approaches such as sustainable land management and ecosystem restoration have proven effective in conserving biodiversity and mitigating climate risks (Mehta et al. 2023).

Despite its benefits, the implementation of NbS still faces several challenges. These include the need for precise planning, comprehensive stakeholder engagement, and ongoing monitoring to ensure that NbS interventions continue to provide their intended benefits. Moreover, the success of NbS often relies on the integration of traditional knowledge with modern scientific practices, ensuring that solutions are both locally relevant and scalable. As such, the widespread adoption of NbS requires a paradigm shift in how we approach environmental management, placing nature at the center of human development strategies (Ruangpan et al. 2019; Debele et al. 2023).

8.3 Environmental Challenges in the Mahanadi River Basin

The Mahanadi River Basin is currently experiencing severe environmental degradation due to a combination of industrial pollution, agricultural runoff, and urban expansion. Industrial effluents, including untreated waste from factories, significantly degrade water quality, impacting both aquatic life and the health of communities that rely on rivers for drinking water and irrigation. The widespread use of chemical fertilizers and pesticides in agriculture has further compounded the problem, leading to eutrophication and a loss of biodiversity in many parts of the basin (Bohorquez et al. 2023).

Soil erosion, driven by deforestation and unregulated land use, poses another major challenge. The clearing of forests for agriculture and urban development has weakened riverbanks, increasing the risk of flooding during the monsoon season. The sedimentation of riverbeds has reduced the capacity of rivers to carry water, exacerbating the risk of both floods and droughts. The combination of these factors has placed immense pressure on basin ecosystems, threatening the livelihoods of millions of people who depend on rivers for their agricultural activities and freshwater resources (Mehta et al. 2023).

Climate change has further intensified these challenges, leading to more frequent and severe weather events such as floods and droughts. Rising temperatures and changing precipitation patterns have disrupted traditional farming practices, making it difficult for farmers to predict planting and harvesting seasons. This has led to reduced crop yields and increased food insecurity in the region. The increasing unpredictability of the climate has made it difficult for local governments to manage water resources effectively, leading to conflicts over water use between different sectors, including agriculture, industry, and urban development (FAO 2021).

Given the scale of these challenges, traditional engineering solutions such as dams and levees are often insufficient, as they fail to address the root causes of environmental degradation. Moreover, these gray infrastructure solutions are expensive to build and maintain and can have negative impacts on ecosystems by disrupting natural water flows and habitats. In contrast, NbS offers a more sustainable and cost-effective alternative that works with nature rather than against it, providing long-term resilience to both environmental and socioeconomic challenges (Ruangpan et al. 2019).

8.4 Potential NbS Strategies for the Mahanadi River Basin

The implementation of nature-based solutions in the Mahanadi River Basin could significantly improve the region's resilience to environmental challenges while enhancing the livelihoods of local communities. Several specific strategies can be applied to the basin, including riparian buffer zones, wetland restoration, sustainable agriculture, and community-led watershed management (Table 8.1).

1. *Riparian Buffer Zones*: Riparian buffer zones are vegetated areas along riverbanks that help reduce soil erosion, filter pollutants, and provide habitats for wildlife. These buffer zones act as natural barriers between the river and agricultural or urban areas, preventing runoff from entering the water and improving water quality. By stabilizing riverbanks, riparian zones can also reduce the risk of flooding and protect downstream communities from the impacts of extreme weather events (Debele et al. 2023).

Table 8.1 Nature-based solutions for Mahanadi River conservation, restoration, and biodiversity protection

Nature-based solution (NbS)	Significance	References
1. Riparian buffer zones	Planting vegetation along riverbanks to prevent soil erosion, filter pollutants, and provide habitats for wildlife	Bohorquez et al. (2023)
2. Wetland restoration	Restoring wetlands to improve water filtration, store floodwater, and enhance biodiversity by creating habitats for aquatic species	Debele et al. (2023)
3. Agroforestry systems	Integrating trees and shrubs into farming systems to reduce soil erosion, sequester carbon, improve biodiversity, and increase water retention	Jose (2009), Debnath et al. (2020)
4. Reforestation	Planting native trees along degraded riverbanks to enhance biodiversity, improve water infiltration, and prevent soil erosion	Palekar (2016), Lal (2020)
5. Sustainable agriculture	Adopting organic farming and crop rotation to reduce chemical runoff, improve soil health, and promote biodiversity	IFOAM (2020), Lal (2020)
6. Green infrastructure	Implementing rain gardens, green roofs, and permeable pavements in urban areas to reduce surface runoff and improve water quality	Ruangpan et al. (2019)
7. Bioengineering for bank stabilization	Using natural materials like plants and biodegradable structures to stabilize riverbanks, reduce erosion, and promote vegetation growth	Palekar (2016)
8. Floodplain restoration	Reconnecting rivers with their natural floodplains to reduce flood risk, improve groundwater recharge, and enhance wildlife habitats	Debele et al. (2023)
9. Conservation agriculture	Implementing no-till farming, cover cropping, and mulching to improve soil structure, prevent erosion, and sequester carbon	Lal (2020), Debnath et al. (2020)
10. Community-based watershed management	Engaging local communities in water resource management and conservation to ensure sustainable practices and riverbank protection	FAO (2021), Palekar (2016)

2. *Wetland Restoration*: Wetlands play a crucial role in regulating water flow and maintaining water quality by acting as natural sponges that absorb excess water during floods and release it slowly during dry periods. Restoring degraded wetlands in the Mahanadi River Basin could help mitigate the impacts of floods and droughts while also providing habitats for diverse plant and animal species. Wetlands are also important for carbon sequestration, as they store large amounts of carbon in their soil and vegetation, making them a key component of climate change mitigation strategies (Ruangpan et al. 2019).
3. *Sustainable Agriculture*: The adoption of sustainable agricultural practices, such as agroforestry, organic farming, and efficient water use, can reduce the negative impacts of agriculture on the environment while improving food security. By reducing the use of chemical fertilizers and pesticides, sustainable farming practices help maintain soil health, prevent water pollution, and increase biodiver-

sity. In the context of the Mahanadi River Basin, promoting sustainable agriculture could help farmers adapt to the changing climate while reducing the environmental footprint of agriculture (Mehta et al. 2023).

4. *Community-led Watershed Management*: Involving local communities in the management of water resources is critical for the success of NbS initiatives. Community-led watershed management programs empower local stakeholders to take ownership of their natural resources, ensuring that conservation efforts are sustainable and tailored to local needs. These programs often involve the restoration of degraded lands, the construction of small-scale water retention structures, and the promotion of water-saving technologies, all of which contribute to improved water security and ecosystem health (FAO 2021).
5. *Green Infrastructure in Urban Areas*: Urban areas within the Mahanadi River Basin contribute to water pollution through runoff from roads, buildings, and other impervious surfaces. Green infrastructure, such as green roofs, rain gardens, and permeable pavements, can help reduce urban runoff and improve water quality by allowing rainwater to infiltrate the soil rather than flow directly into the river. These solutions not only reduce pollution but also increase urban resilience to flooding and heat stress, making cities more liveable and sustainable (Debele et al. 2023).

8.5 Sustainable Farming and Agroforestry Along the Mahanadi River: Enhancing Biodiversity and Reducing Environmental Impacts

The Mahanadi River, one of India's major river systems, serves as a lifeline for millions of people in the states of Chhattisgarh and Odisha. Spanning approximately 858 kilometers, the river is crucial for agricultural irrigation, drinking water, and inland fisheries. However, in recent years, rivers have faced significant environmental challenges, including pollution, soil erosion, loss of biodiversity, and the effects of climate change. The adoption of sustainable agricultural practices, such as natural farming, organic farming, and agroforestry, along riverbanks can play a pivotal role in enhancing biodiversity, reducing environmental impacts, and promoting the conservation of river ecosystems.

The integration of natural and organic farming methods with agroforestry in a 5-kilometer buffer zone along the Mahanadi River has the potential to address environmental degradation while fostering sustainable livelihoods for local communities. These farming systems can mitigate soil erosion, improve water quality, sequester carbon, and support biodiversity by creating habitats for flora and fauna. This study explores how natural farming, organic farming, and agroforestry systems can be implemented along riverbanks to conserve the Mahanadi River and contribute to its ecological and economic health.

8.6 Natural Farming Along the Mahanadi River

Natural farming is an agricultural practice that minimizes human intervention in ecological processes. It is based on the principle that the natural environment has the capacity to sustain itself without the need for synthetic chemicals or fertilizers. This farming method seeks to mimic natural ecosystems by fostering biodiversity, improving soil health, and promoting nutrient cycling (Palekar 2016).

In the context of the Mahanadi River, natural farming can be an effective tool for conservation and biodiversity enhancement. Riverbanks, which have been degraded by conventional agricultural practices, can benefit significantly from the restoration of native vegetation and the reduction in chemical inputs. By encouraging the growth of microorganisms and earthworms, natural farming can improve soil structure and fertility, which, in turn, enhances water retention and reduces soil erosion (Palekar 2016).

One of the key components of natural farming is the use of indigenous microorganisms (IMOs) and vermicompost to replenish soil nutrients. This reduces the need for chemical fertilizers, which are a significant source of pollution in the Mahanadi River. By eliminating synthetic pesticides, natural farming also reduces the risk of water contamination and promotes the growth of beneficial insects, birds, and pollinators (Debnath et al. 2020). This contributes to the overall biodiversity of the river ecosystem, creating a more balanced and resilient environment.

8.7 Organic Farming for Biodiversity and Environmental Protection

Organic farming is another sustainable agricultural practice that can be implemented along the Mahanadi River to protect its ecosystem. Organic farming is characterized by the use of organic fertilizers, crop rotation, green manure, and biological pest control rather than synthetic inputs (IFOAM 2020). This method not only enhances soil fertility but also protects water quality and fosters biodiversity.

By establishing organic farming systems along riverbanks, the harmful effects of chemical runoff from conventional agriculture can be minimized. Chemical fertilizers and pesticides are major contributors to water pollution in rivers, leading to eutrophication and the loss of aquatic biodiversity. By relying on compost, animal manure, and nitrogen-fixing cover crops, organic farming improves soil health and reduces nutrient leaching into rivers (Lal 2020). This is particularly important for the Mahanadi River, which supports a diverse array of aquatic species that are vulnerable to pollution.

In addition to improving soil and water quality, organic farming can play a critical role in enhancing the biodiversity of riverbank ecosystems. Organic farming promotes crop diversity by encouraging the cultivation of a wide variety of crops, including native species that are well adapted to the local environment. This not only reduces the risk of pests and diseases but also creates habitats for a range of organisms, including pollinators and natural predators (Lal 2020). As a result, organic farming contributes to the restoration of ecosystems along riverbanks, helping to create a more stable and diverse environment.

The implementation of organic farming along the Mahanadi River can also contribute to climate change mitigation by sequestering carbon in the soil. Organic matter, such as compost and green manure, helps store carbon in the soil, thus reducing greenhouse gas emissions and increasing the ability of the soil to retain moisture. This makes organic farming a valuable strategy for improving the resilience of river ecosystems to the impacts of climate change (IFOAM 2020).

8.8 Agroforestry Systems for Conservation and Sustainable Livelihoods

Agroforestry, the practice of integrating trees and shrubs into agricultural landscapes, offers a unique opportunity to increase biodiversity, improve soil and water conservation, and promote sustainable livelihoods along the Mahanadi River. Agroforestry systems combine the benefits of agriculture and forestry, creating multifunctional landscapes that provide a range of ecosystem services (Jose 2009). These systems can be designed to reduce soil erosion, improve water infiltration, sequester carbon, and provide habitats for wildlife while supporting productive agricultural activities.

The establishment of agroforestry systems in the 5-kilometer buffer zone along the Mahanadi River can help mitigate the environmental impacts of deforestation and monoculture farming, which have contributed to the degradation of the river ecosystem. Trees and shrubs planted along riverbanks act as natural barriers that reduce soil erosion and filter pollutants from runoff before they enter rivers (Debnath et al. 2020). This helps improve water quality and protects aquatic habitats, which are essential for the survival of fish and other aquatic species.

Agroforestry systems also contribute to biodiversity conservation by creating diverse habitats for a variety of plant and animal species. The inclusion of native tree species in agroforestry systems can provide food and shelter for birds, insects, and mammals, promoting greater biodiversity in the landscape (Jose 2009). Additionally, the integration of fruit trees and medicinal plants into agroforestry systems can provide economic benefits to local communities while preserving traditional knowledge and cultural practices related to the use of these plants.

One of the key advantages of agroforestry is its ability to sequester carbon, making it a valuable tool for climate change mitigation. Trees and shrubs in agroforestry systems capture carbon dioxide from the atmosphere and then store it in their biomass and soil, which helps reduce the overall carbon footprint of agricultural activities (Jose 2009). This is particularly important in the context of the Mahanadi River, where the impacts of climate change, such as more frequent floods and droughts, pose significant risks to local communities and ecosystems.

Agroforestry systems can also improve the resilience of farming systems to climate change by providing shade, windbreaks, and improved soil fertility. The presence of trees in agricultural landscapes helps regulate microclimates, reducing the risk of crop failure due to extreme weather conditions. Additionally, the deep root systems of trees and shrubs increase water infiltration and improve the ability of the soil to retain moisture, making agroforestry systems more resilient to drought (Debnath et al. 2020). By improving the overall health and productivity of the land, agroforestry can help ensure food security and sustainable livelihoods for communities living along the Mahanadi River.

8.9 Integrating Sustainable Practices Along Riverbanks

The integration of natural farming, organic farming, and agroforestry systems along the Mahanadi River offers a holistic approach to river conservation and sustainable development. By promoting biodiversity, improving soil and water quality, and reducing the environmental impacts of conventional agriculture, these practices can help restore the ecological balance of river ecosystems.

One of the most important steps in implementing these sustainable practices is engaging local communities in conservation efforts. Community-based initiatives, such as farmer cooperatives and watershed management groups, can play a crucial role in the successful adoption of natural and organic farming practices along riverbanks. These groups can provide farmers with access to training, resources, and markets for organic and agroforestry products while also fostering a sense of stewardship for the river and its surrounding environment (Palekar 2016).

Government policies and incentives are also essential for promoting the adoption of sustainable agricultural practices. The Indian government's promotion of organic farming through the Paramparagat Krishi Vikas Yojana (PKVY) and other schemes provides financial and technical support to farmers transitioning to chemical-free farming (Lal 2020). By extending these programs to communities along the Mahanadi River, the government can support the large-scale implementation of natural farming, organic farming, and agroforestry systems, contributing to the conservation of the river ecosystem.

The Mahanadi River, with its rich biodiversity and vital role in supporting local communities, is an ecosystem that deserves protection and restoration. The adoption of natural farming, organic farming, and agroforestry systems along riverbanks provides a sustainable pathway for enhancing biodiversity, reducing environmental impacts, and promoting the conservation of river ecosystems. These practices not only improve soil and water quality but also offer economic benefits to local communities by promoting sustainable livelihoods and food security. By integrating these sustainable agricultural practices into the management of the Mahanadi River Basin, we can create a more resilient and biodiverse landscape that supports both human and ecological well-being. The collaboration of local communities, governments, and nongovernmental organizations will be essential in ensuring the long-term success of these conservation efforts.

8.10 Predicted Outcomes of Implementing NbS

The implementation of nature-based solutions in the Mahanadi River Basin is expected to yield a wide range of environmental, social, and economic benefits. By restoring natural ecosystems and promoting sustainable land use practices, NbS can significantly improve water quality, enhance biodiversity, and mitigate the impacts of floods and droughts.

1. *Improved Water Quality*: Riparian buffer zones and wetland restoration are projected to reduce key pollutants such as nitrogen and phosphorus, improving the quality of water for both human use and aquatic ecosystems. This is particularly important for reducing the incidence of waterborne diseases and ensuring the availability of clean water for drinking, agriculture, and industry (Bohorquez et al. 2023).
2. *Enhanced Biodiversity*: By restoring habitats and reducing pollution, NbS can help increase biodiversity in the Mahanadi River Basin. This includes the recovery of fish populations, which are critical for local fisheries, as well as the protection of endangered species that rely on the river and its surrounding ecosystems for survival (Mehta et al. 2023).
3. *Flood Mitigation*: Wetland restoration and the reforestation of degraded lands are expected to reduce the frequency and intensity of floods by absorbing excess water during heavy rains and releasing it slowly during dry periods. This will help protect communities and infrastructure from the devastating impacts of floods, reducing the need for costly flood control measures (Ruangpan et al. 2019).
4. *Increased Agricultural Productivity*: Sustainable agricultural practices, such as agroforestry and organic farming, can improve soil health and water retention, leading to increased crop yields and more resilient farming systems. This is particularly important for ensuring food security in the face of climate change, which is expected to increase the frequency of droughts and other extreme weather events (FAO 2021).
5. *Socioeconomic Benefits*: The socioeconomic benefits of NbS include job creation through restoration projects, increased ecotourism opportunities, and improved livelihoods for farmers and fishermen. By promoting sustainable development, NbS can help alleviate poverty and reduce inequality in the Mahanadi River Basin, contributing to long-term social stability (Debele et al. 2023).

8.11 Conclusion

Nature-based solutions offer a holistic and sustainable approach to addressing the environmental and socioeconomic challenges faced by the Mahanadi River Basin. By working with nature rather than against it, NbS can provide long-term resilience to floods, droughts, and other hydrometeorological risks while enhancing biodiversity and supporting sustainable livelihoods. The successful implementation of NbS in the Mahanadi River Basin will require a combination of scientific knowledge, community engagement, and policy support, but the potential benefits are significant. By restoring ecosystems and promoting sustainable land use practices, NbS can help ensure the continued health and productivity of the Mahanadi River Basin for generations to come.

References

Bohorquez P, Pérez-Latorre FJ, González-Planet I, Jiménez-Melero R (2023) Nature-based solutions for flood mitigation and soil conservation in a steep-slope olive-orchard catchment (Arquillos, SE Spain). Appl Sci 13(2882). https://doi.org/10.3390/app13052882

Debele SE, Leo LS, Kumar P, Sahani J, Ommer J, Bucchignani E (2023) Nature-based solutions can help reduce the impact of natural hazards: a global analysis of NBS case studies. Sci Total Environ 902:165824. https://doi.org/10.1016/j.scitotenv.2023.165824

Debnath B, Pati B, Roy R (2020) Conservation agriculture: an approach towards sustainable farming systems in India. J Clean Prod 263:121344. https://doi.org/10.1016/j.jclepro.2020.121344

FAO (2021) Nature-based solutions in agriculture: The case and pathway for adoption. Virginia: Food and Agriculture Organization of the United Nations and The Nature Conservancy. https://doi.org/10.4060/cb3141en

IFOAM (2020) Organic farming and biodiversity. International Federation of Organic Agriculture Movements. https://www.ifoam.bio/en/organic-policy-guarantee/organic-farming-biodiversity

Jose S (2009) Agroforestry for ecosystem services and environmental benefits: an overview. Agrofor Syst 76(1):1–10. https://doi.org/10.1007/s10457-009-9229-7

Keesstra S, Geissen V, Mosse K, Piiranen S, Scudiero E, Leistra M, Van Schaik L (2018) Soil as a basis to create resilient landscapes. Geoderma 324:70–77. https://doi.org/10.1016/j.geoderma.2018.03.011

Lal R (2020) Organic farming for sustainable agriculture: global perspectives and emerging issues. Adv Agron 163:1–54. https://doi.org/10.1016/bs.agron.2020.07.004

Mehta M, Upgupta S, Singh P, Jha S (2023) Biodiversity conservation in the Hindu Kush Himalayas: a case for nature-based solutions. J Mt Sci 20(3):565–579. https://doi.org/10.1007/s11629-022-7373-8

Palekar S (2016) The philosophy of natural farming. Zero Budget Natural Farming Research Institute

Ruangpan L, Vojinovic Z, Di Sabatino S, Alves A (2019) Nature-based solutions for hydrometeorological risk reduction: a state-of-the-art review of the research area. Sci Total Environ 696:133828. https://doi.org/10.1016/j.scitotenv.2019.133828

9 Land Use/Land Cover Changes Over the Quarter Century in the Paradip Port Region in the Mahanadi Delta of India: An Investigation Using Remote Sensing and Geographical Information System

Tejaswini Dash, Sanjaya Kumar Pattanayak, and Prasanna Kumar

Abstract

Land use/land cover is an important component of understanding the interactions of human activities with the environment, which is required for monitoring. Here, land use/land cover around the Paradip port area was studied through geographical information system (GIS) and remote sensing. The classification was performed using different satellite images, such as Landsat-4 MSS, IRS-1A LISS-II, IRS-P6 LISS-III, and Cartosat-1 PAN, and change detection was studied from 1985 to 2011. The results revealed that approximately 32% of the dense mangrove area has been converted to aquaculture land (8%), sandy areas (5%), scrubland (5%), sparse mangroves (1%), and marshy/swampy land (2%). Rapid changes occurred between 2005 and 2011 because the driving forces were either biophysical or socioeconomic. Biophysical drivers include the natural environment and topographical resources. Socioeconomic drivers include demographic, social, industrial, and technological changes. The driving forces responsible for this phenomenon are overexploitation of fuel, wood, timber harvest, domestic grazing, nontimber forest products, reclamation for agriculture and aquaculture, and the development of rural settlements. Thus, it is vital to focus on the driving factors, and a change detection study is necessary to understand the reasons for the decline in mangrove forests for the sustenance of coastal ecosystems.

T. Dash · S. K. Pattanayak (✉)
P.G. Department of Environmental Sciences, Sambalpur University, Sambalpur, Odisha, India
e-mail: skpattanayak@suniv.ac.in

P. Kumar
Orissa Remote Sensing Application Centre, Bhubaneswar, Odisha, India

J. P. Bhatt, A. K. Mishra (eds.), *Mahanadi River*,
https://doi.org/10.1007/978-981-96-1048-8_9

Keywords

Land use pattern · Change detection · GIS · Remote sensing · Paradip port area

9.1 Introduction

Land use/land cover change (LULC) detection studies through remote sensing and geographic information systems (GISs) provide various types of information for understanding and assessing the phenomenon of the Earth's surface (Olorunfemi 1983). The change detection map shows the different environmental changes and the driving forces responsible for the loss of these factors (Gibson 2000; Gibson and Power 2000). This shift is known to be a significant contributor to environmental change worldwide (Turner et al. 1995). Satellite data obtained via multispectral data and spatial data analysis via GIS can be used to evaluate land use planning and environmental impact in terms of change (Anderson et al. 1976). This information is needed for the environmental management (Adeniyi and Omojola 1999) of such changes from a future perspective (Singh 1989).

It is also critical from the perspective of sustainable development, resource conservation and management of land use planning, climate change modelling, and ecosystem restoration (Shrestha et al. 2022; Jansen and Digregorio. 2004). Evaluation of land cover is essential according to United Nations Global Geodetic Centre of Excellence (UN-GGIM) to protect biodiversity, restore ecosystems, combat climate change, detect shoreline change (Nayak 1986), and manage land resources (Meyer 1995; Dimyati and Kitamura 1996; Di Gregorio and Jansen 1998)..

In this scenario, satellite observations of the surface area are important for providing more information on surface changes. This approach can help us identify land cover. A global land cover pattern was studied by Stanimirova et al. (2023), who examined land use patterns and hotspot areas (Duraisamy 2018) at the surface to understand the changes from the mangrove perspective (Balsamo et al. 2019). ArcGIS Pro, ERDAS, and Google Earth Engine (Sidhu et al. 2018) are the most important techniques for LULC classification (Olaka et al. 2023; Macarringue et al. 2022). This map is very helpful to produce a thematic base map with vector data to predict and gather all the information. Geographically referenced data can be captured, stored, queried, analyzed, and displayed with the aid of GIS (Good Child 2000; Wilkinson 1996).

With the help of satellite data and thematic base map data, soil moisture estimation crop yield forecasting, LULC, change detection (Trabelsi et al. 2023; Abebe et al. 2021; Civco et al. 2002; Baynard 2013), coastal wetland mapping (Rebelo et al. 2009; Zhou et al. 2023), change detection (Hurd et al. 1992), forest and urban sprawl mapping (Hurd et al. 2001; Liao et al. 2008), and ecosystem restoration and land use planning and mangrove and coastal vegetation mapping

(Nayak and Bahugunam 2001; Kuenzer et al. 2011; Sunkur et al. 2024) can be achieved. The most popular satellites are Landsat, SPOT, and IRS (Lillesand and Kiefer 1994).

Change detection in land use/cover can be performed on a temporal scale, such as a decadal scale, to assess landscape changes caused by anthropogenic activities on land (Gibson and Power 2000). These anthropogenic activities on land are the result of rapid urbanization and industrialization. It has also been recognized as an important driver of global environmental change (Turner et al. 1995; Wang et al. 2023). Xiaomei and Rongqing (1999) reported that changing information is necessary for updating land cover maps and managing natural resources. Currently, neural networks and AI-based techniques are the best alternatives in this field (Civco 1993; Wang and Civco 1996). This helps us understand object-based remote sensing land cover classification, the change correlation matrix, the RGB-NDVI, the confusion matrix (Liao et al. 2008), and accuracy assessment (Wang and Civco 1992).

Several studies have effectively evaluated the dynamics of LULC and their determinants at the local level. Many of these investigations have made use of commercial geographic analysis programs such as ArcGIS for visualizing the Earth Resources Data Analysis System (ERDAS) and images (ENVI), among others, for mapping and LULC change detection (Hurd et al. 2001).

In the Mahanadi Basin, the land use/land cover of the delta region has undergone drastic changes for a period of 25 years, spanning from 1985 to 2011, particularly in the Paradip port region. This has affected the mangrove forest of the region; mangroves are important landcover areas of the coastal tract whose decline has adverse impacts on coastal ecosystems. Therefore, this study highlights the land use/land cover changes around the Paradip port area to understand how the mangrove area changed from 1985 to 2011 and to ascertain the factors responsible for its decline.

9.2 Materials and Methods

9.2.1 Study Area

The coast of Paradip is one of the most important coastal regions where several coastal-based developments have taken place in the last 20 years. It is spread between 20°15′0″ to 21°50′0″ N latitude and 86°30′0″ to 87°10′0″ E longitude (Fig. 9.1). It falls in the Mahanadi River Basin covering the Jagatsinghpur and Kendrapara districts in the state of Odisha. Mangrove swamps and a network of creeks are prevalent in the region. In addition, the area is important because it houses a port near the estuary, one fertilizer plant, one brewery, and other fishing-related

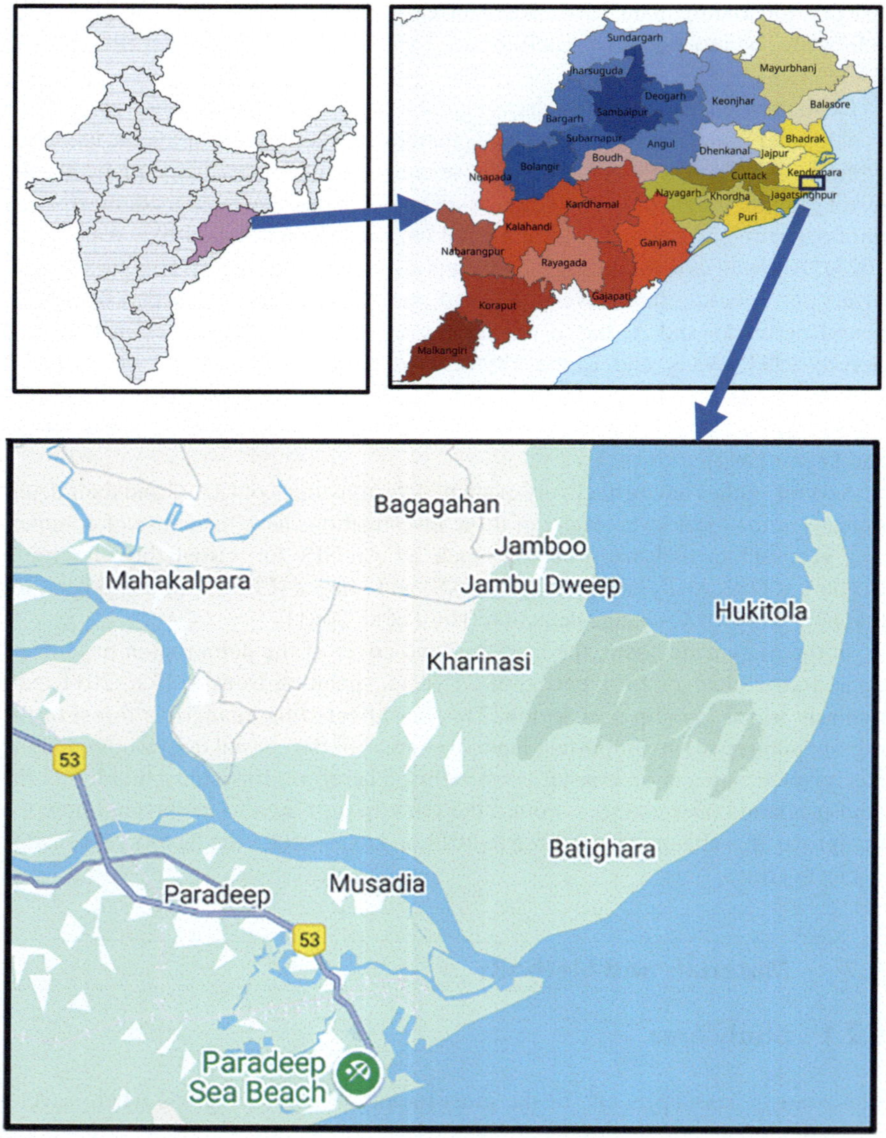

Fig. 9.1 The location map of the study area (Region around the Paradip port area)

industries. Thus, the area offers an excellent scope for studying geomorphic aspects and is important from the point of view of environmental management in terms of conservation and development.

For this present study, multispectral multitemporal satellite data from the Paradip port area were acquired for four temporal points, viz., 1985, 1995, 2005, and 2011.

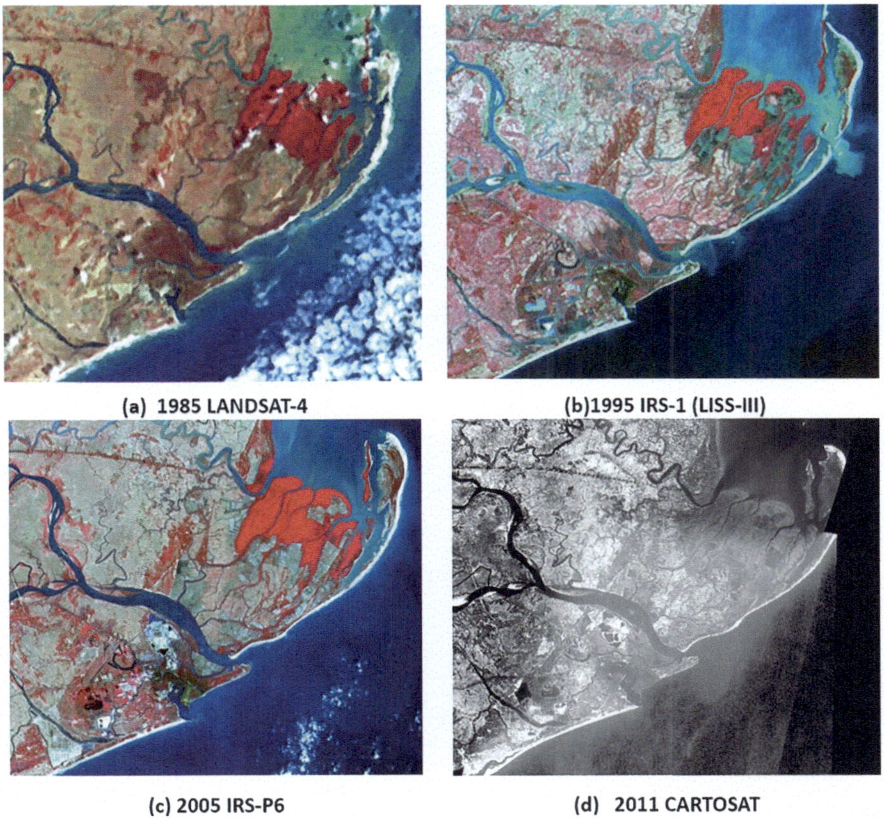

Fig. 9.2 Satellite pictures for the study region (**a**) 1985 LANDSAT-4 data, (**b**) 1995 IRS-1 data, (**c**) 2005 IRS-P6 data, (**d**) 2011 CARTOSAT data

The 1985 image is a Landsat MSS image with a resolution of 79 meters. While the IRS images from 1995 and 2011 had resolutions of 36 m and 23 m, respectively. The Cartosat image from 2005 had a spatial resolution of 5.8 m. These maps were registered via the Universal Transverse Mercator (UTM) projection zone 45 N. All the satellite images are shown in Fig. 9.2. Using the following methodology, land use/land cover maps were prepared from these four satellite images, and the changes in land use/land cover were determined.

9.3 Methodology

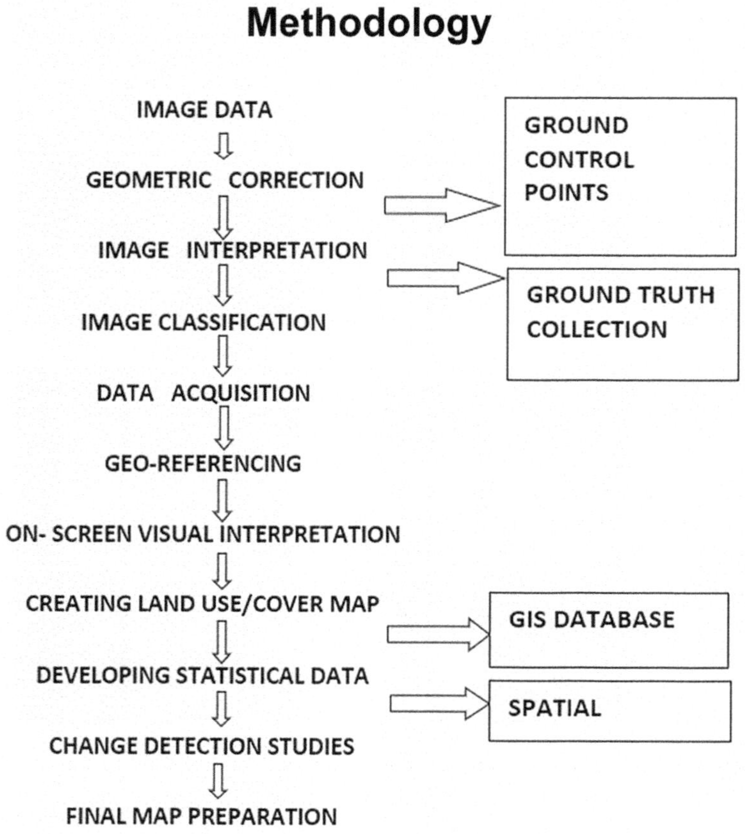

9.4 Results and Discussion

The land use/land cover maps were prepared via four different satellite images, which are shown in Fig. 9.2. Then, the change detection analysis results were compared for the four periods given in Fig. 9.3. The changed land use/land cover maps from 1985 to 2011 are shown in Fig. 9.4. The change detection matrix shows the change in land use/land cover data from 1985 to 1995 in Table 9.1.

This change detection analysis revealed that 57% of the ocean area was converted into mudflat areas and that 40% of the ocean area was converted to sandy areas from 1985 to 1995. There was very little change (5%) in settlement areas, whereas cropland changed into industrial areas in 1995. A major portion of the mangrove area was converted to mudflats (13%) and sparse mangroves (2%). Among the total area, 83% of the area remained unchanged from 1985 to 1995. Those areas where sparse mangroves were present in 1985 changed into croplands (12%), degraded mangroves (29%), mudflats (5%), marshy swampy lands (4%),

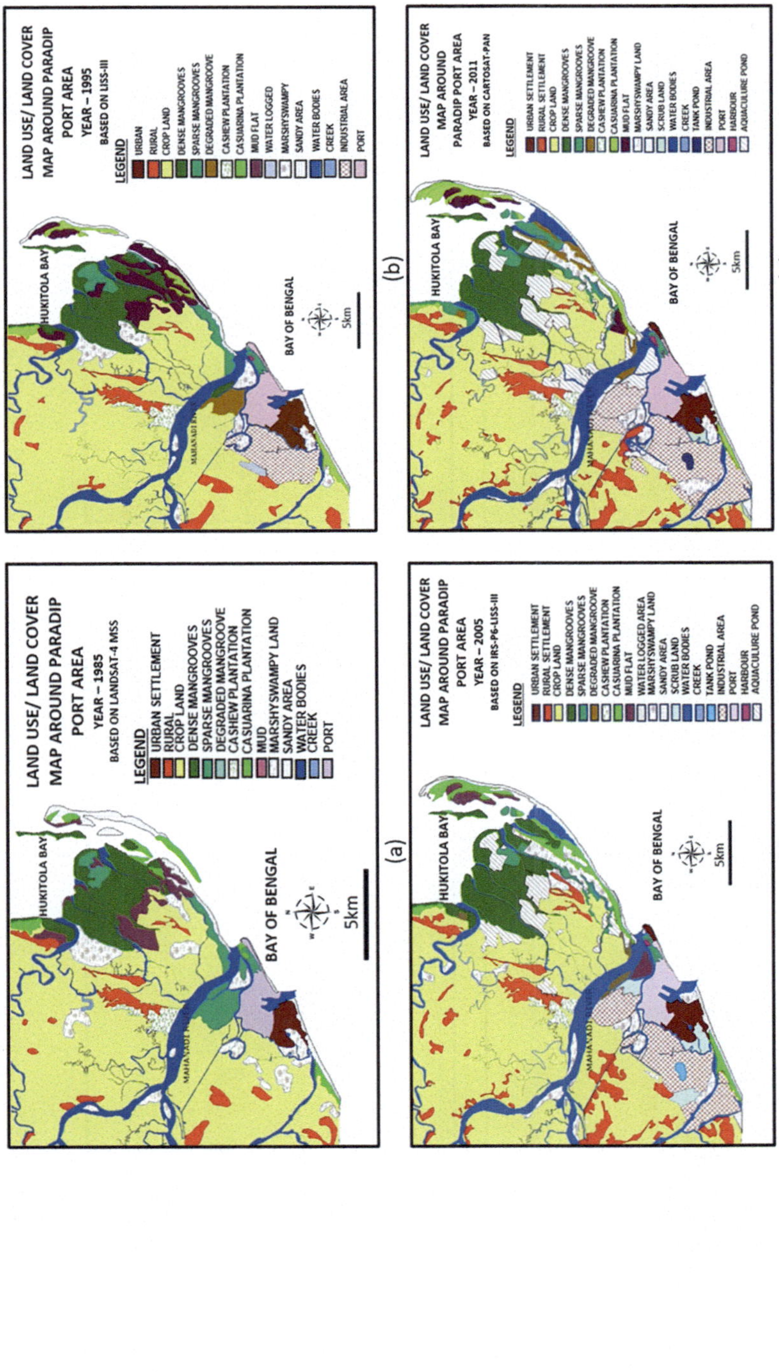

Fig. 9.3 Land use/land cover map of the Paradip port area **(a)** 1985, **(b)** 1995, **(c)** 2005, and **(d)** 2011

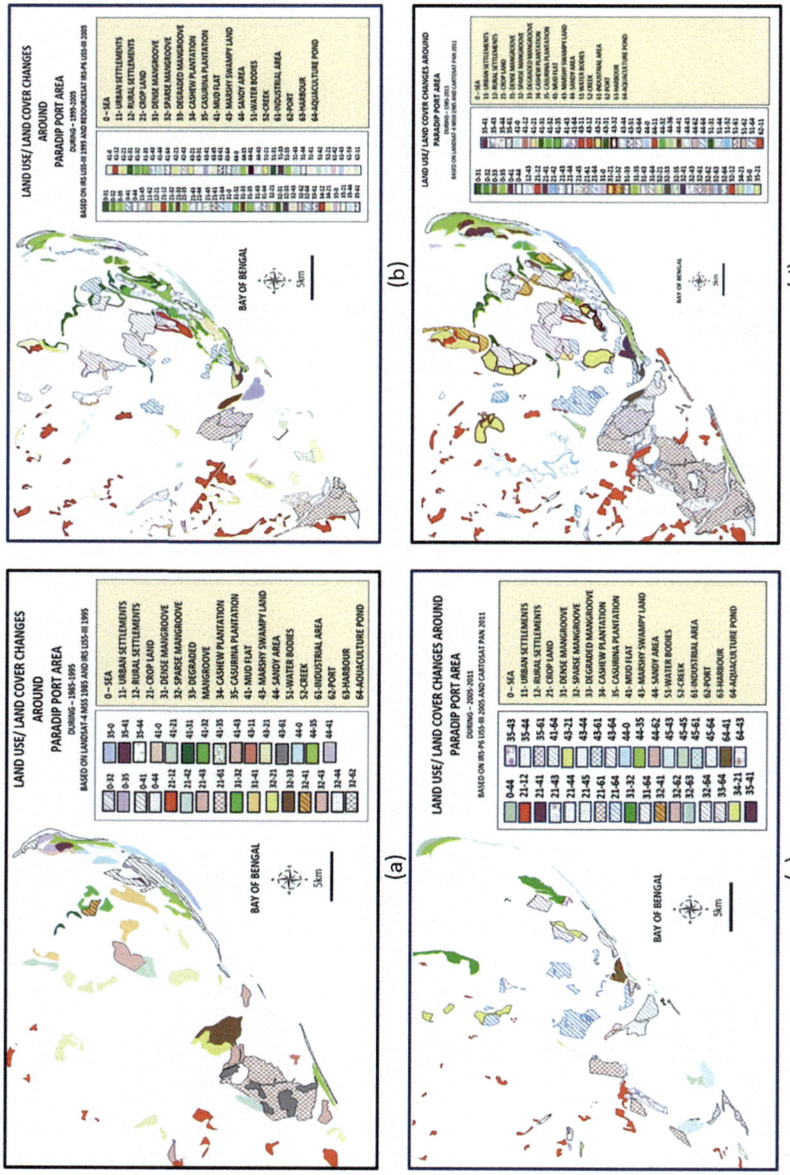

Fig. 9.4 Change detection in land use/land cover around the Paradip port area from **(a)** 1985–1995, **(b)** 1995–2005, **(c)** 2005–2011, **(d)** 1985–2011

Table 9.1 The change detection matrix showing the change in land cover/land use from 1985 to 1995

	0	11	12	21	31	32	33	34	35	41	42	43	44	45	51	52	53	61	62	63	64	Total
0						8.04 (0.81%)			16.02 (1.61%)	571 (57%)			398.64 (40%)									993.7 (2.3%)
11		**693** **(99%)**											4.22 (0.60%)									697.23 (1.70%)
12			**1602** **(100%)**																			1602.2 (3.90%)
21			416.79 (1%)	**21034** **(90%)**		15.96				12.02 (0.05%)	119.46 (0.52%)	267.9 (1%)			74.32 (0.32%)	32.71 (0.14%)		1184.5 (5%)				23157.79 56.41%
31	17.1 (0.52%)				**2783** **(83%)**	78.33 (2%)				435.63 (13%)												3313.9 (8.07%)
32				204.56 (12%)	82.95 (5%)	**660.3** **(40%)**	483.22 (29%)			82.79 (5%)		64.65 (4%)	13.12 (0.81%)						38.04 (2%)			1629.7 (3.97%)
33																						
34								**385.8** **(92%)**							29.94 (7%)							415.72 (1.01%)
35	175.46 (23%)								**431.7** **(57%)**	76.59 (10%)			9.48 (1.25%)		49.08 (6%)					13.39 (1%)		755.72 (1.84%)
41	65.89 (5%)			238.86 (19%)	46.32 (3%)	37.05 (3%)			3.21 (0.26%)	**561.9** **(41%)**		262.61 (21%)			5.1 (0.42%)							1220.95 (2.97%)
42																						
43		27.47 (1%)		926.73 (42%)						53.75 (2%)		**703.7** **(32%)**			47.33 (2%)	21.03 (0.96%)		418.38 (19%)				2198.35 (5.35%)
44	276.44 (25%)								396.44 (36%)	109.14 (10%)			**305.6** **(28%)**						2.62 (0.24%)			1090.2 (2.66%)
45																						
51				46.75 (1%)								66.88 (1%)	15.84 (0.39%)		**3892** **(96%)**					0.02		4021 9.79%
52				21.03 (41%)												**29.77** **(58%)**						50.8 0.12%
62		52.48 (5%)																	**847.4** **(94%)**			899.89 2.19%
Total	**534.88**	**772.96**	**2019.01**	**22472.06**	**2912.1**	**799.71**	**483.22**	**385.82**	**847.34**	**1902.85**	**119.46**	**1365.7**	**746.91**		**4097.31**	**83.51**		**1602.88**	**889.07**	**13.41**		**42047.29**
	1.27%	**1.86%**	**4.86%**	**54.13%**	**7.02%**	**1.93%**	**1.16%**	**0.93%**	**2.04%**	**4.58%**	**0.29%**	**3.29%**	**1.80%**	**0.00**	**9.87%**	**0.20%**	**0.00**	**3.86%**	**2.14%**	**0.03**	**0.00**	

Table 9.2 The change detection matrix showing the change in land cover/land use from 1995 to 2005

	0	11	12	21	31	32	33	34	35	41	42	43	44	45	51	52	53	61	62	63	64	Total
0					99.69 (18%)	59.67 (11%)			69.85 (13%)	87.85 (16%)			217.83 (40%)									**534.89 1.27%**
11		**692 (89%)**												68.29 (8%)					12.62 (1%)			**772.97 1.86%**
12			**1866 (92%)**	141.55 (7%)											3.71 (0.18%)			7.82 (0.39%)				**2019.01 4.86%**
21			1216.1 (5%)	**19,409 (86%)**					2.91 (0.01%)			439.95 (1.96%)	48.79 (0.22%)	162.04 (0.72%)				886.32 3%)			306.73 (1%)	**22472.1 54.13%**
31	38.04 (1%)				**2346.08 (80%)**	96.81 (3%)	272.92 (9%)		123.41 (4%)			26.13 (0.90%)			2.83						5.92 (0.20%)	**2912.14 7.02%**
32		2.06 (0.26%)		39.97 (5%)	38.54 4%	**239.1 (29%)**	152.8 (19%)		90.5 (11%)			122.44 (15%)			5.41 (0.68%)				8.69 (1%)		100.18 (12%)	**799.71 1.93%**
33							**123.17 (25%)**							110.08 (22%)	3.74 (0.77%)			246.22 (50%)				**483.21 1.16%**
34			35.99 (9%)	35.91 (9%)				**313.92 (81%)**														**385.82 0.93%**
35	33.11 (3%)			88.36 (10%)					**696.35 (82%)**			1.81 (0.21%)	9.46 (1%)					18.3 (2%)				**847.39 2.04%**
41	29.28 (1%)		6.38 (0.34%)	59.83 (3%)	163.65 (8%)	62.74 (3%)			67.29 (3%)	**171.8 (9%)**		570.16 (29%)	98.39 (5%)		253.6 (13%)						419.74 (22%)	**1902.85 4.58%**
42				81.5 (68%)								28.82 (24%)			9.14 (7%)							**119.46 0.29%**
43				67.57 (4%)	23.92 (1%)					178.6 (13%)		**329.5 (24%)**		117.4 (8%)	17.97 (1%)			92.57 (6%)			538.17 (39%)	**1365.7 3.29%**
44	223.2 29%	17.3 2%							124.12 (16%)	12.14 (1%)		28.01 (3%)	**317 (42%)**		15.2 (2%)				9.91 (1%)			**746.91 1.80%**

45																						
51				57.46 (1%)	72.56 (1%)				4.23 (0.10%)			48.41 (1%)	12.28 (0.30%)		**3700** **(90%)**	175.11 (4%)		14.12 (0.34%)	11.28 (0.28%)	1.86 (0.05%)		**4097.31** **9.87%**
52				62.52 (74%)												**20.99** **(25%)**						**83.51** **0.20%**
53																						
61											20.14 (1.26%)	2.7 (0.17%)					74.29 (4%)	**1505.8** **(93%)**				**1602.88** **3.86%**
62		90.07 (10%)																	**798** **(89%)**			**888.07** **2.14%**
63																				**13.41** **(100%)**		**13.41** **0.03%**
64																						
Total	**323.66** **0.77%**	**801.49** **1.92%**	**3124.40** **7.49%**	**20043.89** **48.04%**	**2744.44** **6.58%**	**458.33** **1.10%**	**548.90** **1.32%**	**313.92** **0.75%**	**1178.66** **2.82%**	**450.41** **1.08%**	**20.14** **0.05%**	**1597.92** **3.83%**	**703.75** **1.69%**	**457.81** **1.10%**	**4011.58** **9.61%**	**196.10** **0.46%**	**74.29** **0.18%**	**2771.10** **6.64%**	**840.50** **2.02%**	**15.27** **0.04%**	**1370.64** **3.29%**	**42047.30**

and port areas (2%). Among the plantation area, 92% remained unchanged from 1985 to 1995. Among the cashew plantation areas, 57% remained unchanged, 7% changed to river and stream areas, and 10% converted into mudflat areas. The area in the mudflat category changed to cropland (19%), mangrove (3%), sparse mangrove (3%), and marshy/swampy land (21%). Of the mudflat area, 46% remained unchanged. Among the marshy/swampy land, 42% changed to cropland, 19% changed to industry, and 32% remained unchanged in 1995. Among the area covered with sand in 1985, 36% changed to casuarina plantations, while 28% remained unchanged in 1995. The water body and port areas remained unchanged from 1985 to 1995 (Table 9.1).

The changing patterns in land use/land cover from 1995 to 2005 are given in Table 9.2. The change detection analysis revealed that the ocean area was converted into mangrove forest (3%), sparse mangrove (11%), casuarina plantation (35%), and sandy areas (44%). Among the urban areas, 89% remained unchanged. Among the rural areas, 92% remained unchanged, and 7% changed to cropland. The area covered by cropland in 1995 changed to settlements (5%) and industry (3%); however, 86% of the area remained unchanged. Dense mangrove areas were converted into degraded mangroves (9%), but 80% remained unchanged from 1995 to 2005. A major portion of the sparse mangrove area was converted to cropland (5%), degraded mangrove (19%), cashew plantation (11%), marshy swampy land (15%), and aquaculture land (12%). However, 29% of this area remained unchanged from 1995 to 2005.

Of the degraded mangrove area, 25% remained unchanged from 1995 to 2005, and it was mostly converted into scrubland (22%). Cashew plantations changed to rural settlements (9%) and croplands (9%) in 2005, but casuarina plantations were converted to croplands (10%), industrial areas (2%), and sandy areas (1%). A major portion of the mudflat area changed to cropland (3%), dense mangrove (8%), sparse mangrove (3%), casuarina plantation (3%), marshy swampy land (29%), sandy area (5%), water bodies (13%), and aquaculture land (22%). Sixty-eight percent of the waterlogged area changed to cropland, and 24% changed to marshy swampy land from 1995 to 2005. Marshy swampy land changed to cropland (4%), mudflat (13%), scrubland (8%), industrial areas (6%), and aquaculture land (39%) from 1995 to 2005. Among the water bodies, 90% remained unchanged from 1995 to 2005, but 4% of the area changed to creek. There was a slight change (7%) in the industrial area. There was no aquaculture land in 1995, but water bodies were converted to aquaculture land by 0.05% in 2005.

The changes in land use/cover from 200 to 2011 are given in Table 9.3, which shows that 323 ha of land were converted from ocean areas to sand bodies from 2005 to 2011. The major portion of sparse mangroves (67%) remained unchanged, and the remaining part was converted to degraded mangroves (2%), mudflats (4%), harbor areas (2%), and aquaculture land (3%). Among the degraded mangroves, 90% remained unchanged, but only 9% were converted into aquaculture. Among the mudflat area, 60% remained unchanged, but 39% of the area was converted to aquaculture. There was no change in the waterlogged area. Marshy swampy land

Table 9.3 The change detection matrix showing the change in land cover/land use from 2005 to 2011

	0	11	12	21	31	32	33	34	35	41	42	43	44	45	51	52	53	61	62	63	64	Total
0													323.66 (100%)									**323.66 0.77%**
11		**757 (94%)**																	44.8 (5%)			**801.49 (1.92%)**
12			**3120.6 (99%)**																		3.8 (0.12%)	**3124.40 (7.49%)**
21			366.07 (1%)	**18,097 (90%)**								179.42 (1%)	52.91 (0.26%)	25.36				538.68 (2%)			784.6 (3%)	**20043.48 (48.04%)**
31					**2418.3 (88%)**	321.81 (11%)															4.34 (0.16%)	**2744.44 (6.58%)**
32						**311.54 (67%)**	98.91 (21%)			19.41 (4%)				0.59					2.03	11.87 (2%)	13.97 (3%)	**458.33 (1%)**
33							**496.9 (90%)**														52.01 (9%)	**548.91 (1%)**
34				23 (7%)				**290.92 (92%)**														**313.91 (0.75%)**
35									**976 (82%)**	5.72		49.62 (4%)	46.33 (3%)					101.01 (8%)				**1178.65 (2%)**
41										**272 (60%)**											178.61 39%	**450.43 (1%)**
42											**20.14 (100%)**											**20.14 (0.05%)**
43		1.06		233.94 (14%)			1.63		8.32			**947.4 (59%)**	1.13		44.21 (2%)			43.21 (2%)			317.02 19%	**1597.94 (3%)**
44	91.5 (13%)	6.52							103.16 (14%)				**475 (67%)**		21.87 (3%)				5.73			**704.05 (1%)**
45														**315.4 (68%)**				5.73 (1%)			136.71 (29%)	**457.81 (1%)**
51				25.46								0.33	32.4		**3953 (98%)**							**4011.43 (9%)**

(continued)

Table 9.3 (continued)

	0	11	12	21	31	32	33	34	35	41	42	43	44	45	51	52	53	61	62	63	64	Total
52				8.01 (4%)											37.05 (18%)	**151.05 (77%)**						**196.11 (0.47%)**
53																	74.29 (100%)					**74.29 0.18%**
61			59.21 (2%)									17.24			17.15			**2639.2 (95%)**			38.34 (1%)	**2771.10 6.64%**
62		2.82				7.51 (0.89%)													**830.17 (98%)**			**840.90 (2.02%)**
63																				**15.27 (100%)**		**15.27 0.04%**
64												190.7 (13%)				9.51					**1171 (85%)**	**1370.64 (3%)**
Total	**91.5 0.21%**	**767.08 (1.83%)**	**3545.88 (8.45%)**	**18387.24 (43.83%)**	**2418.30 (5.76%)**	**641.26 1%**	**597.44 (1.42%)**	**290.92 (0.69%)**	**1087.44 (2.54%)**	**296.93 0.71%**	**20.14 (0.05%)**	**1384.70 3.30%**	**931.40 2.22%**	**340.73 0.81%**	**4083.67 9.73%**	**151.05 0.36%**	**74.29 (0.18%)**	**3327.83 7.93%**	**883.03 (2%)**	**27.14 0.06%**	**2699.43 6.43%**	**42047.28**

Table 9.4 The change detection matrix showing the change in land cover/land use from 1985 to 2011

	0	11	12	21	31	32	33	34	35	41	42	43	44	45	51	52	53	61	62	63	64	Total
0					86.50 (8%)	75.55 (7%)	16.92 (1%)		299.99 (30%)	18.54 (1%)			297.3 (29%)		198.82 (20%)							**993.70 (2.3%)**
11		**630. (90%)**												67.10 (9%)								**697.23 1.07%**
12			**1307.5 (81%)**	233.02 (14%)								3.52 (0.22%)		58.14 (3%)								**1602.24 3.90%**
21			2133.1 (9%)	**17118.8 (73%)**		86.82 (0.37%)				42.51 (0.18%	20.1	725.20 (3%)	58.60	20.36	276.84 (1%)	5.65	5.37	2055.0 (8%)			609.29 (2%)	**23157.79 56.41%**
31	20.4 (0.62%)			9.34 (0.28%)	**2155.6 (65%)**	192.2 (5%)	56.09 (1%)		11.23			86.68 (2%)	191.0 (5%)	195.1 (5%)	113.85 (3%)	2.00			2.00	0.12	278.16 (8%)	**3313.94 8.07%**
32				4.07		**172.4 (10%)**	212.3 (13%)		21.91 (1%)	65.58 (41%)		3.39			19.84 (1%)			402.26 (24%)	2.85	11.8	713.13 (43%)	**1629.67 3.97**
33																						
34			35.99 (8%)	50.95 (12%)				**290.9 (69%)**							37.90 (9%)							**415.76 1.01%**
35	21.40 (2%)			79.96 (10%)		21.62 (2%)	171.3 (22%)		**258.37 (34%)**	57.81 (7%)		4.08	5.34		20.59 (2%)			100.02 (13%)		15.1 (2%)		**755.72 1.84%**
41	26.38 (2%)		4.86	197.38 (27%)	77.03 (6%)	13.44 (1%)	19.92 (1%)		7.11	**29.49 2%**		184.78 (15%)			5.80			153.00 (12%)			501.77 (41%)	**1220.95 2.97%**
42																						
43		25.46 (1%)	59.51 (2%)	597.46 (27%)	23.92 (1%)	33.04 (1%)						**309.23 (14%)**	124.0 (5%)		80.42 (3%)		68.9 (3%)	389.71 (17%)			486.90 (22%)	**2198.62 5.35%**
44	23.29 (2%)	23.82 (2%)				45.15 (4%)	117.0 (10%)		381.77 (35%)	31.01 (2%)		12.30 (1%)	**199.2 (18%)**		9.77			213.64 (19%)	25.23 (2%)		8.12	**1090.36 2.66%**
45																						
51			4.81	61.88 (1%)	75.20 (1%)		3.74		107.05 (2%)	51.99 (1%)		55.56 (1%)	55.91 (1%)		**3295.6 (81%)**	128.5 (3%)		14.20	64.65 (1%)		102.05 (2%)	**4021.31 9.79%**
52				34.30												**15.80**						**50.11 0.12%**

(continued)

Table 9.4 (continued)

	0	11	12	21	31	32	33	34	35	41	42	43	44	45	51	52	53	61	62	63	64	Total
53																						
61																						
62		87.64 (9%)													24.25 (2%)				**788.0 (87%)**			**899.89 2.19%**
63																						
64																						
Total	**91.46 0.21%**	**767.0 1%**	**3545.8 8.45%**	**18387.2 43.83%**	**2418.3 5.76%**	**640.3 1%**	**597.4 1.4%**	**290.9 0.61%**	**1087.4 2.59%**	**296.9 0.71%**	**20.1 0.05%**	**1384.7 3.30%**	**931.4 2.22%**	**340.7 0.81%**	**4083.7 9.63%**	**152.0 0.36%**	**74.3 0.18%**	**3327.8 7.93%**	**882.7 2.10%**	**27.1 0.06%**	**2699.4 6.43%**	**42047.28**

0, ocean; 11, urban settlements, 12; rural settlements; 21, crop land; 31, dense mangrove; 32, sparse mangrove; 33, degraded mangrove; 34, cashew plantation; 35, casuarina plantation; 41, mud flat; 42, waterlogged area; 43, marshy swampy land; 44, sandy area; 45, scrub land; 51, river; 52, creek; 53, tank/pond; 61, industrial area; 62, ports; 63, harbor; and 64, aquaculture pond

was converted to cropland (14%), industrial areas (2%), and aquaculture land (19%). The areas that covered the sandy area changed into ocean areas (13%), casuarina plantations (14%), and water bodies (3%). For the total area, a large chunk of land (29%) was converted into aquaculture land. Water bodies remained unchanged from 2005 to 2011. The major portion of aquaculture land (85%) remained unchanged, but only 13% changed to marshy swampy land.

The changes in land use and land cover from 1985 to 2011 are shown in Table 9.4. This indicates that ocean areas were converted to dense mangroves (8%), sparse mangroves (7%), casuarinas (30%), mud flats (41%), sandy areas (29%), and river and water bodies (20%). Among the scrublands, 90% remained unchanged, but the majority of the cropland changed into settlements (9%), marshy/swampy land (3%), rivers (1%), industrial areas (8%), and aquaculture ponds (2%). Dense mangrove areas were converted to sparse mangrove areas (5%), degraded mangrove areas (1%), marshy swampy land areas (2%), sandy areas (5%), scrubland areas (5%), river water bodies (8%), and aquaculture ponds (8%). Only 65% of the areas remain unchanged. The areas under sparse mangroves were converted into degraded mangroves (13%), casuarina plantations (1%), mudflats (4%), industrial areas (24%), and aquaculture ponds (43%), and only 10% remained unchanged. Among cashew plantations, 69% of the area remained unchanged in 2011, and the rest were converted to settlements (8%), croplands (12%), or rivers (9%). Camellia plantations were converted into croplands (10%), mudflats (7%), and industrial areas (6%). The areas covered with mudflats changed to croplands (6%), dense mangroves (6%), marshy/swampy lands (15%), industrial areas (12%), and aquaculture ponds (41%). Marshy land was converted into settlements (2%), croplands (27%), sandy areas (5%), industrial areas (17%), and aquaculture ponds (22%) in 2011. In the case of river and water bodies, 81% remained unchanged in 2011, and the remainder was converted to casuarina plantations and aquaculture ponds (2%). Among the port area, 87% remained unchanged, but some portions (9%) changed to urban settlements.

9.5 Discussion

This study of the multitemporal and multispectral satellite data for the region in and around Paradip port provides crucial information concerning the changes in land use and land cover patterns of the region that have taken place over the last 25 years from 2005 to 2011. The mangrove ecosystem of the region is affected. In the dense mangrove region, 32% of the area has been converted to different land uses/land covers. A related study carried out by Mani Murali et al. (2004) on Paradip port area using satellite IRS-1D and IRS-III data revealed a significant reduction in mangroves and an increase in urban land, which was attributed to human interference in the natural environment. Another study on the Gujrat coast by Nayak and Bahuguna (2001) revealed significant changes in the mangrove and coral reefs during the period 1975–1985 by using IRS satellite data.

In the Paradip port area, major conversion of the original land use/land cover of the region has been to aquaculture land (8%), sandy areas (5%), scrubland (5%), sparse mangroves (5%), degraded mangroves (1%), and marshy/swampy land (2%). Rapid changes (21% conversion) from mangroves to sparse mangroves occurred between 2005 and 2011. Aquaculture activity seems to be an important cause of the degradation of mangrove forests in the region. Marshy swampy land, sandy areas, and the majority of the port area have been converted to urban settlements. These shifts were at a maximum between 1995 and 2005. Rural settlements have developed on cropland (9%), cashew plantation areas (8%), and marshy/swampy land (2%). The conversion of cashew plantation areas to rural settlements was maximal from 1995 to 2005. There has been a large-scale transformation of the casuarina plantation area and mud flat. Approximately 40% of the area of the mud flat region has been converted to aquaculture land. In 22% of the Casuarinas plantation area, few mangroves have developed.

Compared with natural activities, the growth of settlements and aquaculture activities have intensive effects on the dynamics of land use/land cover changes in a region. Zubair (2006) utilized remote sensing and GIS technologies to detect land use and/or land cover changes in Ilorin, Nigeria, from 1972 to 2001, through Landsat TM images via the supervised classification method to delineate five land use/cover classes for this study. They reported that farmland declined by 7%, whereas built-up land increased by 8%. Land use/land cover change (LULC) disturbs the structure and function of ecosystems and thereby perturbs their functionality, decreasing the ability of the ecosystem to provide resources to humans (Kesgin and Nurlu 2009; Fu et al. 2015; Hasan et al. 2020; Tonyaloğlu 2024). LULC results from any process, such as population growth, economic development, or agricultural demand, that adversely impacts landscapes, climate change, biodiversity, and thereby human health and well-being (Verburg et al. 2013; Petroni et al. 2022; Tonyaloğlu 2024).

9.6 Conclusion

Land use change is easily discernable from multitemporal data. The changes depict the changing environmental conditions of the area as well as giving recognition to the affected different drivers for over 26 years. The natural drivers play fewer roles than the anthropological drivers do. Associated industrial activity and anthropogenic dependence move toward an urbanized setup that significantly impacts the estuarine area. Mangrove ecosystems are regarded as vital biological resources. However, their degradation affects biodiversity. Therefore, the changes in the mangrove ecosystem in the Mahanadi delta region in the vicinity of Paradip need continual mapping via high-resolution satellite imagery and monitoring to maintain a sustainable ecosystem in the region.

Acknowledgement TD acknowledges the support extended by ORSAC and Odisha to conduct this study.

References

Abebe G, Getachew D, Ewunetu A (2021) Analysing land use/land cover changes and its dynamics via remote sensing and GIS in Gubalafito district, Northeastern Ethiopia. SN Appl Sci 4(30):1–15. https://doi.org/10.1007/s42452-021-04915-8

Adeniyi PO, Omojola A (1999) Land use landcover change evaluation in Sokoto—Rima Basin of North Western Nigeria based on archival of the environment (AARSE). In: Geoinformation technology applications for resource and environmental management in Africa, pp 143–172

Anderson JR, Hardy EE, Roach JT Wither RE (1976) A Land use and Land cover classification System for use with remote sensor data. USDI, Geol. Survey Professional Paper 964, Washington

Balsamo G, Agustì-parareda A, Albergel C, Arduini G, Beljaars A, Bidlot J, Blyth E, Bousserez N, Boussetta S, Brown A, Buizza R, Buontempo C, Chevallier F, Choulga M, Cloke H, Cronin MF, Dahoui M, De Rosnay P, Dirmeyer PA, Zeng X (2019) Satellite and in situ observations for advancing global earth surface modelling: a review. Remote Sens 11(8):941. https://doi.org/10.3390/rs11080941

Baynard CW (2013) Remote sensing applications : beyond land-use and land-cover change. Adv Remote Sensing 2:228–241. https://doi.org/10.4236/ars.2013.23025

Civco DL (1993) Artificial neural networks for land cover classification and mapping. Int J Geogr Inf Syst 7(2):173–186. https://doi.org/10.1080/02693799308901949

Civco D, Hurd JD, Wilson EH (2002) A comparison of land use and land cover change detection methods, 1–13. https://www.researchgate.net/publication/228543190

Di Gregorio A, Jansen LJM (1998) A new concept for a land cover classification system. Land 2(1):55–65

Dimyati MUM, Kitamura T (1996) An analysis of land use/cover change using the combination of MSS Landsat and land use map: a case study in Yogyakarta, Indonesia. Int J Remote Sens 17(5):931–944. https://doi.org/10.1080/01431169608949056

Duraisamy V (2018) Identifying hotspots in land use land cover change and the drivers in a semiarid region of India. Environ Monit Assess:190–535. https://doi.org/10.1007/s10661-018-6919-5

Fu B, Zhang L, Xu Z, Zhao Y, Wei Y, Skinner D (2015) Ecosystem services in changing land use. J Soils Sediments. https://doi.org/10.1007/s11368-015-1082-x

Gibson PJ (2000) Introductory remote sensing. Principles and concepts with contributions from Clare H. Power and John Keating. Routledge, London, p 184

Gibson PJ, Power CH (2000) Introductory remote sensing. Digital image processing and applications. Routledge. Geological Magazine 138(02):219–232, London, p 249. https://doi.org/10.1017/S0016756801244951

Good Child MF (2000) Communicating geographic information in a digital age. Ann Assoc Am Geogr 90(2):344–355. https://doi.org/10.1111/0004-5608.00198

Hasan SS, Zhen L, Miah MG, Ahamed T, Samie A (2020) Impact of land use change on ecosystem services: a review. Environ Dev. https://doi.org/10.1016/j.envdev.2020.100527

Hurd JD, Civco DL La Bash C, August P (1992) Coastal wetland mapping and change detection in the Northeast United States. In: Proceedings of 1992 ASPRS/ACSM/RT'92 convention. Washington, vol 1, p 130–139

Hurd JD, Wilson EH, Lammey S, Civco DL (2001) Characterization of forest fragmentation and urban sprawl using time sequential Landsat imagery. In: Proceedings of 2001 ASPRS annual convention, St. Louis, MO, p 13. https://www.researchgate.net/publication/228703481

Jansen LJM, Digregorio. (2004) Obtaining land use information from a remotely sensed land cover map: results from a case study in Lebanon. Int Jour Appl Earth Obs Geoinf 5:141–157. https://doi.org/10.1016/j.jag.2004.02.001

Kesgin B, Nurlu E (2009) Land cover changes on the coastal zone of Candarli Bay. EMAS, Turkey using remotely sensed data. https://doi.org/10.1007/s10661-008-0517-x

Kuenzer C, Bluemel A, Gebhardt S, Quoc TV, Dech S (2011) Remote sensing of mangrove ecosystems: a review. Remote Sens 3:878–928. https://doi.org/10.3390/rs3050878

Liao M, Jiang L, Lin H, Huang B, Gong J (2008) Urban change detection based on coherence and intensity characteristics of SAR imagery. Photogramm Eng Remote Sens 74(8):999–1006

Lillesand T, Kiefer R (1994) Remote sensing and image interpretation, 3rd edn. Wiley, New York, p 750

Macarringue LS, Bolfe ÉL, Roberto P, Pereira M (2022) Developments in land use and land cover classification techniques in remote sensing : a review. J Geogr Inf Syst 14:1–28. https://doi.org/10.4236/jgis.2022.141001

Mani Murali R, Shakuntala Ekka P, Venthamony D (2004) Langovan has worked on Land use/cover studies of areas in and around Paradip Port, East Coast, p 441–445. https://www.researchgate.net/publication/27667486

Meyer WB (1995) Past and present land use and land-cover in the U.S.A. Consequences: the nature and implications of environmental change, p 24–33

Nayak S, Bahugunam A (2001) Application of remote sensing data to monitor mangroves and other coastal vegetation of India. Ind J Mar Sci 30(4):195–213

Nayak SR (1986) Remote sensing of shoreline changes in the Mahi estuary. In the proceedings of the national seminar on remote sensing of water resources, IWRS and ISPI and RS, Ahmedabad, *2*, p 30–38

Olaka LA, Elhag MM, Adan M, Tonnang HEZ (2023) Mapping, intensities, and future prediction of land use/land cover dynamics using Google earth engine and CA- artificial neural network model. PLoS One 1–28. https://doi.org/10.1371/journal.pone.0288694

Olorunfemi JF (1983) Monitoring urban land – use in developed countries – an aerial photographic approach. Environ Int 9:27–32. https://doi.org/10.1016/0160-4120(83)90111-3

Petroni ML, Siqueira-Gay J, Gallardo ALCF (2022) Understanding land use change impacts on ecosystem services within urban protected areas. Landsc Urban Plan 223:104404. https://doi.org/10.1016/j.landurbplan.2022.104404

Rebelo LM, Finlayson CM, Nagabhatla N (2009) Remote sensing and GIS for wetland inventory, mapping and change analysis. J Environ Manag 90:2144–2153. https://doi.org/10.1016/j.jenvman.2007.06.027

Shrestha S, Poudyal KN, Bhattarai N, Dangi MB (2022) An assessment of the impact of land use and land cover change on the degradation of ecosystem service values in Kathmandu Valley using remote sensing and GIS. Sustain For 14(23):15739. https://doi.org/10.3390/su142315739

Sidhu N, Pebesma E, Câmara G (2018) Using Google earth engine to detect land cover change: Singapore as a use case using Google earth engine to detect land cover change: Singapore as a use. Eur J Remote Sens 51(1):486–500. https://doi.org/10.1080/22797254.2018.1451782

Singh A (1989) Digital change detection techniques using remotely sensed data. Int J Remote Sens 10(6):989–1003. https://doi.org/10.1080/01431168908903939

Stanimirova R, Tarrio K, Turlej K, McAvoy K, Stonebrook S, Hu KT, Friedl MA (2023) A global land cover training dataset from 1984 to 2020. Sci Data 10(1):879

Sunkur R, Kantamaneni K, Bokhoree C, Rathnayake U, Fernando M (2024) Mangrove mapping and monitoring using remote sensing techniques towards climate change resilience. Sci Rep:1–14. https://doi.org/10.1038/s41598-024-57563-4

Trabelsi N, Triki I, Hentati I, Rachdi N (2023) Classification and Spatio-temporal change detection of land use/land cover using remote sensing and geographic information system in the Manouba region. NE Tunisia:652–668. https://doi.org/10.4236/jgis.2023.156033

Tonyaloğlu EE (2024) Future land use/land cover and its impacts on ecosystem services: case of Aydın, Turkey. Int J Environ Sci Technol. https://doi.org/10.1007/s13762-024-05907-y

Turner, B., Skole, D., Sanderson, S., Fisher, G., Fresco, L. & Leemans, R. (1995). Land use and land cover change science/research plan. International human dimensions of global environmental change programme (IHDP) Report No. 7, p 20–22

Verburg PH, Erb KH, Mertz O, Espindola G (2013) Land system science: between global challenges and local realities. Curr Opin Environ Sustain. https://doi.org/10.1016/j.cosust.2013.08.001

Wang Q, Xiong K, Zhou J, Xiao H (2023) Impact of land use and land cover change on the landscape pattern and service value of the village ecosystem in the karst desertification control. Front Environ Sci:1–15. https://doi.org/10.3389/fenvs.2023.1020331

Wang Y, Civco DL (1992) Evidential reasoning and postclassification of misclassified pixels: a GIS approach for improving the classification accuracy of remotely sensed data. In: Proceedings of 1992 ASPRS/ACSM/RT'92 convention, vol 4, Washington, pp 160–170

Wang Y, Civco DL (1996) Three artificial neural network paradigms in high dimensional multisource spatial data classification. Geogr Inf Sci 1(2):73–87. https://doi.org/10.1080/10824009509480473

Wilkinson GG (1996) A review of current issues in the integration of GIS and remote sensing. Int J Geogr Infor Sys 10:85–101

Xiaomei Y, Ronqing LQY, (1999). Change detection based on remote sensing information model and its application to coastal line of Yellow River Delta Earth Observation Center, NASDA, China, p 107–121. https://www.geospatialworld.net/article

Zhou X, Wang J, Zheng F, Wang H, Yang H (2023) An overview of coastline extraction from remote sensing data. Remote Sens 2023(15):4865. https://doi.org/10.3390/rs15194865

Zubair A (2006) Change detection in land use and land cover using remote sensing data and GIS: case study of Ilorin and its environs in Kwara State. GIS Dev J

Risk Assessment of Heavy Metals from the Ib and Bheden Rivers: The Main Tributaries of Mahanadi in Western Odisha, India

10

Malaya Ranjan Mahananda, Bidut Prava Mohanty, and Rakesh Roshan Sahu

Abstract

This study investigated the presence of heavy metals in fish, water, and soil near the Jharsuguda district, along the Ib and Bheden rivers. Water samples were collected between January and June 2023, and the results revealed varying levels of cadmium, chromium, and lead in the water. Sediments show seasonal and average fluctuations in these metals, with lead having the highest average quantity. The decreasing order of metal accumulation in the sediments was Pb > cr > cd. The fish samples presented monthly concentrations of cadmium, chromium, and lead in the muscle of *Labeo rohita*, with cadmium being the most accumulated metal in the fish. The bioconcentration factor (BCF) for heavy metals is highest for lead, followed by chromium and cadmium. The average BCF values for lead, chromium, and cadmium are 10.7142, 0.0055, and 5.513 L/kg, respectively. This study also analyzed the monthly concentrations of cadmium, chromium, and lead in the muscle of *Mystus gulio* and *Sardinella fimbriata*. Lead was the most accumulated metal in the fish, whereas cadmium was the least accumulated one. The BCF of heavy metals in the muscle of *Sardinella fimbriata* was noted to be similar, with chromium having the highest BCF. This study emphasizes the importance of addressing these contaminants in fish health and reducing their impact on aquatic ecosystems.

Keywords

Bioconcentration factor · Heavy metals · BCF · Fish · Sediments · Water

M. R. Mahananda (✉) · R. R. Sahu
P.G. Department of Environmental Sciences, Sambalpur University, Sambalpur, Odisha, India
e-mail: malaya_env@suniv.ac.in

B. P. Mohanty
P.G. Department of Zoology, Panchayat College, Bargarh, Odisha, India

J. P. Bhatt, A. K. Mishra (eds.), *Mahanadi River*,
https://doi.org/10.1007/978-981-96-1048-8_10

10.1 Background

Pollution is a global issue that significantly impacts people's health, with the effects of pollution from automobiles, industrial effluents, and congested cities being alarming (Khan and Ghouri 2011; Ali et al. 2016; Hashim et al. 2014; Gowd et al. 2010). The environment, plants, and health of people and animals all suffer significantly from excessive pollution (Begum et al. 2009). Pollution affects 20 million children globally and has worsened due to population growth (Khan and Ghouri 2011). Over 1.2 billion people are thought to drink dirty water, which is the cause of illnesses that kill between five and ten million people worldwide, the majority of whom are children (Kanmony 2009; Giri and Singh 2014).

Heavy metals are the most prevalent environmental pollutants in the world (Balali-Mood et al. 2021; Ghorade et al. 2014), and their presence can have toxic effects on the human population (Ahuja 2009; Simeonov et al. 2010). Environmental monitoring prioritizes the understanding of heavy metal accumulation in soils, their sources, and potential interactions with soil properties (Al-Weher 2008; Kumar and Dua 2009). Owing to possible health risks, concerns about food safety, and negative impacts on soil ecosystems, the accumulation of heavy metals in agricultural soils is a growing concern (Sarkar 2005; Ahmad et al. 2021).

Long-term exposure to heavy metals can cause them to accumulate in the soil at toxic levels. The application of fertilizers and untreated wastewater can lead to the accumulation of heavy metals in surface soils that are irrigated by wastewater (Gupta et al. 2009; Khan and Ghouri 2011). When this capacity is diminished by repeated wastewater application, heavy metals leak into groundwater or soil solutions available for plant uptake (Papatilippaki et al. 2008). Research shows that applying large amounts of fertilizer can contaminate groundwater by allowing nitrates and heavy metals to leach into the ground, which can then lower the water quality (Fong and Molnar 2008; Gabrielyan et al. 2018; Siziba et al. 2021).

Challenges linked to long-term exposure to heavy metals include severe brain and hematological damage, anemia, and kidney malfunction (Papatilippaki et al. 2008). Lead (Pb) and copper (Cu) are examples of heavy metals that can be fatal in minuscule amounts, whereas exposure to cadmium (Cd) can result in renal dysfunction, abnormalities related to calcium metabolism, and an increased risk of certain types of cancer (Qishlaqi and Moore 2007; Prasad et al. 2020; Proshad et al. 2021; Bhat et al. 2023). Chromium compound exposure has been linked to skin ulcers, malignant neoplasia, and bronchial asthma (Mico et al. 2006). Manganese (Mn) poisoning can impact the central nervous system, hand steadiness, visual reaction time, and eye–hand coordination (Voet et al. 2008). A condition known as manganism is typified by tremors, a feeling of weakness, lethargic mask-like countenance, and mental instability. Chronic inhalation exposure at work has also been linked to respiratory effects in workers. Thus, the main objective of this present study is to measure the concentrations of heavy metals in the fish, water, and soil near the Jharsuguda district along the Ib and Bheden rivers and assess their health risks.

10.2 Research Design and Site Description

This study used a nonprobability sampling design to select cases that contributed most to the research. Two rivers, Ib and Bheden, in the Jharsuguda district, were chosen because of their proximity to large-scale mining operations, industries, construction, and improper wastewater disposal. The rivers flow into the Hirakud Reservoir when they join the Mahanadi River, and numerous industries along the banks discharge wastewater into the river (Fig. 10.1).

The Ib River Basin (21°34′N 83°48′E) in northeastern Central India is a rich coal belt that drains 12,447 square kilometers and travels approximately 252 kilometers. The Ib Valley Coalfield is home to the majority of Mahanadi Coalfields and has experienced the growth of numerous industries. The Bheden River, originating in the Bonai subdivision of Odisha, flows through the districts of Sambalpur, Kuchinda, and Laida before joining Ib and ending up in the Hirakud Dam Reservoir. The non-monsoon season is a marginal time for the river.

In the Bheden River (21°46′7″N 84°14′40″E) watershed, there is one integrated iron and steel industry, one thermal power plant, and one aluminum smelter plant. All industries have installed effluent treatment plants to handle their industrial wastewater. Thermal power plants have implemented the recirculation of ash pond effluent and the zero discharge concept to reduce water pollution. The only smelter plant in the Bheden River catchment has a 5040 KLD effluent treatment plant to

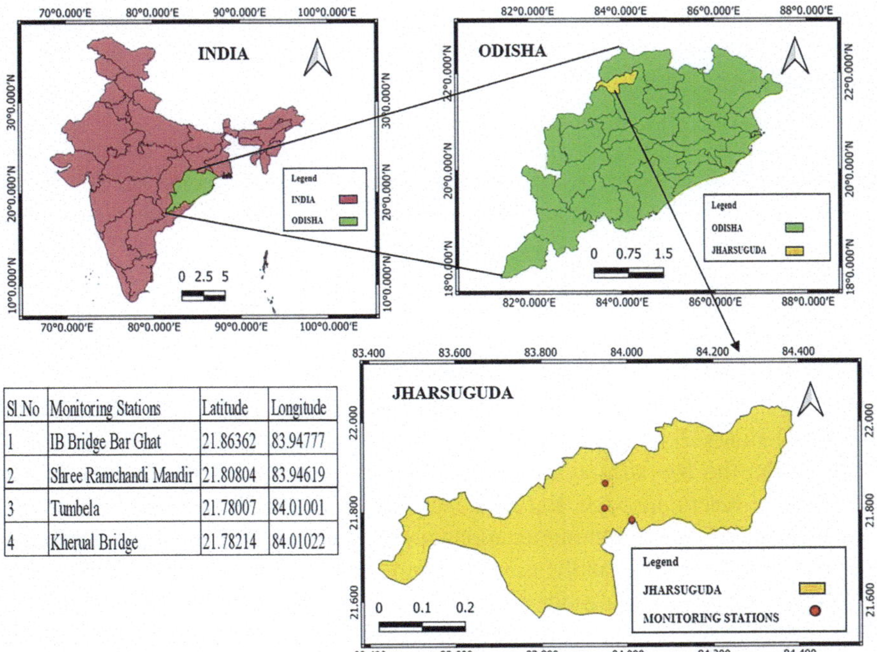

Sl No	Monitoring Stations	Latitude	Longitude
1	IB Bridge Bar Ghat	21.86362	83.94777
2	Shree Ramchandi Mandir	21.80804	83.94619
3	Tumbela	21.78007	84.01001
4	Kherual Bridge	21.78214	84.01022

Fig. 10.1 Sampling points of the study area

treat industrial wastewater before it is released into Kharkhari Nala. The sector has set up two stations for continuous effluent quality monitoring.

Domestic Wastewater Jharsuguda, the largest urban local body in the Bheden River catchment, and three industrial townships have implemented sewage treatment plants to limit wastewater discharge to adjacent water bodies. Approximately 3.12 MLDs of wastewater are treated in these townships by captive wastewater treatment plants (STPs). Jharsuguda stores wastewater in ponds or low-lying areas, but overflow is discharged into the Ib River because of the area's natural slope, reducing the risk of household wastewater contamination in the Bheden River.

Industrial Waste The Bheden River watershed is primarily a waste area. The ash produced by thermal power plants dispose of their ash in slurry form in designated ash pond areas, implementing the Ministry of Fly Ash Utilization Notification by the Ministry of Environment, Forest and Climate Change. Ash silos are installed to store fly ash for future uses, such as in ash-based products, soil conditioners, coal mine void filling, and low-lying areas. Additionally, hazardous waste from various industries is disposed of in secure landfills, ensuring that the water quality of the Bheden River remains unaffected by solid or hazardous waste disposal.

Municipal Solid Waste The Jharsuguda municipality and industrial establishments dispose of their municipal solid waste in designated areas that are located far from river embankments. Consequently, the river is unaffected by the dumping of municipal solid waste.

Biomedical Waste There is just one non-bedded dispensary in the Bheden River catchment. Therefore, the amount of biomedical waste generated in this catchment area is negligible, and there is a remote chance that biomedical waste in the area could contaminate the Bheden River.

10.3 Collection of Water Samples

Between January 2023 and June 2023, water samples were taken once every *two* months from the *two sites* at the three monitoring stations on the Bheden River, including the Kherual Bridge, Bar Gate, Tumbela, Maa Ramchandi Mandir (mixing point), and Ib Bridge beginning locations in the direction of Brajarajnagar and the bridge terminus. To preserve the metals and prevent precipitation, the samples were taken using a Nansen-type water sampler from a depth of 1 foot below the surface. They were then stored in polythene containers (500 mL), with the addition of 2 milliliters of concentrated HNO3. A pH meter with an electronic glass electrode was used to measure the conductivity and pH of the water samples.

10.4 Collection of Sediment Samples

As shown in Fig. 10.1, five benthic sediment samples were collected from the study zone of the Bheden and Ib rivers in a sequential upstream to downstream manner. In January 2023 and after utilization of a shovel, monthly benthic sediment samples were taken from the river between January 2023 and June 2023 (Simpson and Batley 2016) at various sample points during a 6-month period (January–February, March–April, May– June). By wading the river from the upstream side three times, samples were taken from the riverbed when the water depth was approximately 1 meter (USEPA 2020).

10.5 Collection of Fish Samples

Five sexually mature fish species associated with high consumption by the local population were collected from local fishermen and different dominant local markets at each study site. In total, five samples representing the five species from the study site were collected and wrapped in polyethylene bags; then, the samples were transported on ice to the laboratory for biometrics, dissection, and collection of fish tissue for heavy metal analysis.

10.6 Sample Preparation

10.6.1 Digestion Procedure for the Water Sample

The milestone acid assimilation method was used to absorb the metal. Each water sample was pipetted into a 20 ml Teflon tube in 5 milliliters. Each sample received the addition of concentrated acids, i.e., 6 ml HNO3, 65%; 3 ml HCl, 37%; and 0.25 ml H_2O_2. The samples spent 30 min in the ETHOS 900 microwave digester. Upon absorption, the samples underwent cooling. The VARIAN AA240FS Fast Sequential AAS was then used to analyze the heavy metal concentrations in the fluid using the specified instrument parameters.

10.6.2 Digestion Procedure for the Sediment Samples

In total, 1.5 g of sediment sample was weighed into a 20 ml Teflon tube, and each sample was then treated with strong acids (6 ml HNO3, 65%; 3 ml HCl, 37%; and 0.25 ml H_2O_2). After 30 min, the sample was placed in an ETHOS 900 microwave digester and cooled. Each sample was filtered via a pipe fitted with Whatman no. 41 channel paper and placed into a 20 ml Teflon tube after being mixed with 5 ml of refined water. The VARIAN AA240FS Fast Sequential AAS was then used to analyze the concentrations of heavy metals in the fluid via the specified instrument parameters.

10.6.3 Digestion Procedure for the Fish Samples

The fish samples were dried at 105 °C for 24 h after being cleaned with distilled water. The fish samples were dried, and the muscle, head, tail, eyes, and gills were kept for analysis, while the bones and scales were removed. Before being digested, they were ground using a mortar and pestle. A mixture of 2 g of ground fish sample, 5 ml of HNO3, and 2 ml of HCLO4 was heated to 85 °C for 30 min on a hot plate. The samples that had been digested were cooled and then placed in a 50 ml volumetric tube with distilled water added. AAS was used to analyze the filtrate. The wavelengths at which the analyses were conducted were 228.8 nm for cadmium, 357.9 nm for chromium, and 283.3 nm for lead.

10.6.4 Bioconcentration Factor (BCF)

Aquatic toxicology uses the word "bioconcentration" to describe this process. It makes it possible to determine whether a substance's concentration within a living thing is greater than that of its surroundings. There is a connection with the bioconcentration factor, which is the ratio of the concentrations of heavy metals in water to those in living things. The ratio of an organism's heavy metal concentration to that of water defines this component (Kennish et al. 1992). It has the following relationship and is expressed in L/kg (Kennish et al. 1992):

$$\text{BCF} = \text{C}_{\text{MF}} / \text{C}_{\text{MW}}$$

where CMW represents the concentration of heavy metals in the water and CMF represents the concentration in the fish muscle.

10.7 Results and Discussion

The amounts of chromium, cadmium, and lead observed in the river water samples are displayed in Table 10.1. The cadmium values ranged between 0.002 and 0.039 mg/L. The levels of chromium vary from 0.006 to 0.015 mg/L, whereas the lead concentration ranges between 1.511 and 2.044 mg/L.

The seasonal and average fluctuations in the lead, chromium, and cadmium metal concentrations in the sediments are displayed in Table 10.2. Lead has the highest average quantity in sediments, followed by chromium and cadmium. The decreasing order of metal accumulation in the sediments was Pb > cr > cd. The lead, chromium, and cadmium concentrations in the sediments are higher than those in the water. Lead has the greatest average concentration in sediments, followed by cadmium and chromium. The cadmium content in the sediment ranged from 0.02 to 0.04 mg/kg. The chromium content ranged from 0.01 to 0.02 mg/kg, and the lead content ranged from 1.2 to 2.35 mg/kg.

Table 10.1 Heavy metal concentrations in water by month

	Water sample	Cd (mg/l)	Cr(mg/l)	Pb (mg/l)
Jan–Feb	A1	0.041	0.011	2.044
	A2	0.025	0.098	1.925
	A3	0.033	0.108	1.932
	A4	0.037	0.137	1.929
	A5	0.039	0.033	1.963
Mar–Apr	B1	0.028	0.067	1.931
	B2	0.024	0.092	1.993
	B3	0.025	0.052	1.982
	B4	0.023	0.040	1.975
	B5	0.021	0.08	1.984
May–June	C1	0.02	0.021	1.543
	C2	0.021	0.022	1.53
	C3	0.031	0.019	1.511
	C4	0.03	0.017	1.672
	C5	0.029	0.023	1.672
Mean		0.028457	0.0524	1.835
SD		0.006696125	0.041625542	0.197123964

Table 10.2 Monthly heavy metal concentrations in the sediments

	Sediment sample	Cd (mg/kg)	Cr (mg/kg)	Pb (mg/kg)
Jan–Feb	A1	0.04	0.212	2.351
	A2	0.035	0.207	2.195
	A3	0.039	0.088	2.137
	A4	0.035	0.234	1.981
	A5	0.038	0.127	1.824
Mar–Apr	B1	0.034	0.134	1.679
	B2	0.045	0.146	1.876
	B3	0.035	0.163	1.92
	B4	0.051	0.141	1.346
	B5	0.023	0.166	1.689
May–June	C1	0.034	0.153	1.875
	C2	0.034	0.153	1.875
	C3	0.024	0.132	1.823
	C4	0.032	0.151	1.369
	C5	0.39	0.14	1.245
		0.043	0.127	1.345
Mean		0.059866667	0.154733333	1.777
SD		0.091610875	0.037727343	0.333682484

10.8 Fish Sample Preparation

Table 10.3 shows the monthly concentrations of cadmium, chromium, and lead found in the muscle of *Labeo rohita* collected from the river water. These values range from 0.143 to 0.159 mg/kg for cadmium, with a mean value of 0.15 mg/kg;

Table 10.3 Bioconcentration (mg/kg) of heavy metals in the muscle of *Labeo rohita*

Water sample	Cd (muscle	Cd (water)	BCF	Cr (muscle)	Cr (water)	BCF	Pb (muscle)	Pb (water)	BCF
AW1	0.159	0.041	3.878049	0.255	0.011	0.010572	2.087	2.044	1.040452
AW2	0.159	0.025	6.36	0.255	0.098	0.003931	2.087	1.925	24.9312
AW3	0.159	0.033	4.818182	0.255	0.108	0.006849	2.087	1.932	15.7686
AW4	0.159	0.037	4.297297	0.255	0.137	0.00861	2.087	1.929	15.91161
AW5	0.159	0.039	4.076923	0.255	0.033	0.009566	2.087	1.963	3.449704
BW1	0.143	0.028	5.107143	0.22	0.067	0.005483	1.674	1.931	12.22066
BW2	0.143	0.024	5.958333	0.22	0.092	0.004028	1.674	1.993	22.84028
BW3	0.143	0.025	5.72	0.22	0.052	0.004371	1.674	1.982	11.8976
BW4	0.143	0.023	6.217391	0.22	0.006	0.003699	1.674	1.975	1.621928
BW5	0.143	0.021	6.809524	0.22	0.08	0.003084	1.674	1.984	25.94104
CW1	0.149	0.02	7.45	0.261	0.021	0.002685	1.453	1.543	7.8225
CW2	0.149	0.021	7.095238	0.261	0.022	0.00296	1.453	1.53	7.433107
CW3	0.149	0.031	4.806452	0.261	0.019	0.00645	1.453	1.511	2.94589
CW4	0.149	0.03	4.966667	0.261	0.017	0.00604	1.453	1.611	2.814444
CW5	0.149	0.029	5.137931	0.261	0.023	0.005644	1.453	1.672	4.074911
Mean	0.150		5.513275	0.245333		0.005598	1.738		10.71426
Std. dev	0.006		1.108098	0.018715		0.002452	0.271978		8.669469

from 0.22 to 0.261 mg/kg for chromium, with a mean value of 0.24 mg/kg; and between 1.453 and 2.087 mg/kg for lead, with a mean value 1.738 mg/kg. Lead was the most accumulated metal in the fish during the entire sampling period, whereas cadmium was the least accumulated heavy metal.

The bioconcentration factor (BCF) for heavy metals: Lead has the highest BCF, according to previous studies, followed by chromium and cadmium. The average BCF values for lead, chromium, and cadmium are 10.7142, 0.0055, and 5.513 L/kg, respectively. During the dry season, the bioconcentration factor is noted to be greater. The BCF (Pb) > BCF (Cd) > BCF (Cr) ranking indicates the decreasing order of the metal bioconcentration factor. This arrangement corresponds to the sequence in which heavy metals accumulate in fish muscle.

Table 10.4 shows the monthly concentrations of cadmium, chromium, and lead found in the muscle of *Oreochromis niloticus* collected from the water. These values range from 0.04 to 0.08 mg/kg for cadmium; from 0.211 to 0.303 mg/kg for chromium; and between 1.671 and 2 mg/kg for lead. Lead was the most accumulated metal in the fish during the entire sampling period, whereas cadmium was the least accumulated one. Cadmium and chromium are the most abundant metals during the dry season.

The bioconcentration factor (BCF) for heavy metals: Cadmium has the highest BCF, according to previous studies, followed by chromium and lead. The average BCF values for lead, chromium, and cadmium are 0.008714, 0.016312, and 2.427416 L/kg, respectively. During the dry season, the bioconcentration factor is noted to be greater. The BCF (Cd) > BCF (Cr) > BCF (Pb) ranking indicates a decreasing order of the metal bioconcentration factor. This arrangement corresponds to the sequence in which heavy metals accumulate in fish muscle (Table 10.5).

The table shows the monthly concentrations of cadmium, chromium, and lead found in the muscle of *O. cotio* collected from the water. These values range from 0.019 to 0.024 mg/kg for cadmium; from 0.212 to 0.238 mg/kg for chromium; and from 1.873 to 1.953 mg/kg for lead. Lead was the most accumulated metal in this fish during the entire sampling period, whereas cadmium was the least accumulated one. Cadmium and chromium are the most abundant metals during the dry season.

The bioconcentration factor (BCF) for heavy metals. Chromium has the highest BCF, according to previous studies, followed by chromium and lead. The average BCF values for lead, chromium, and cadmium are 1.816982, 3.44934, and 0.776621 L/kg, respectively. During the dry season, the bioconcentration factor is greater. The BCF (Cr) > BCF (Pb) > BCF (Cd) ranking indicates a decreasing order of the metal bioconcentration factor. This arrangement corresponds to the sequence in which heavy metals accumulate in fish muscle.

Table 10.6 shows the monthly concentrations of cadmium, chromium and lead found in the muscle of *Mystus gulio* collected from the water. These values range from 0.013 to 0.016 mg/kg for cadmium; from 0.174 to 0.185 mg/kg for chromium; and between 1.743 and 1.955 mg/kg for lead. Lead was the most accumulated metal in the fish during the entire sampling period, whereas cadmium was the least accumulated one. Cadmium and chromium are the most abundant metals during the dry season.

Table 10.4 Bioconcentration (mg/kg) of heavy metals in the muscle of *O. niloticus*

Water sample	Cd (muscle)	Cd (water)	BCF	Cr (water)	Cr (muscle)	BCF	Pb (water)	Pb (muscle	BCF
AW1	0.04	0.041	0.97561	0.011	0.238	0.042025	2.044	2	0.02056
AW2	0.04	0.025	1.6	0.098	0.238	0.015625	1.925	2	0.008117
AW3	0.04	0.033	1.212121	0.108	0.238	0.027225	1.932	2	0.014092
AW4	0.04	0.037	1.081081	0.137	0.238	0.034225	1.929	2	0.017742
AW5	0.04	0.039	1.025641	0.033	0.238	0.038025	1.963	2	0.019371
BW1	0.08	0.028	2.857143	0.067	0.303	0.0098	1.931	1.671	0.005075
BW2	0.08	0.024	3.333333	0.092	0.303	0.0072	1.993	1.671	0.003613
BW3	0.08	0.025	3.2	0.052	0.303	0.007813	1.982	1.671	0.003942
BW4	0.08	0.023	3.478261	0.006	0.303	0.006613	1.975	1.671	0.003348
BW5	0.08	0.021	3.809524	0.08	0.303	0.005513	1.984	1.671	0.002778
CW1	0.07	0.02	3.5	0.021	0.211	0.005714	1.543	1.715	0.003703
CW2	0.07	0.021	3.333333	0.022	0.211	0.0063	1.53	1.715	0.004118
CW3	0.07	0.031	2.258065	0.019	0.211	0.013729	1.511	1.715	0.009086
CW4	0.07	0.03	2.333333	0.017	0.211	0.012857	1.611	1.715	0.007981
CW5	0.07	0.029	2.413793	0.023	0.211	0.012014	1.672	1.715	0.007186
Mean			2.427416			0.016312			0.008714
Std. dev.	0.017		1.023735		0.038	0.012624		0.145	0.006211

Table 10.5 Bioconcentrations (mg/kg) of heavy metals in the muscle of *Osteobrama cotio*

Samples	Cd (muscle)	Cd (water)	BCF	Cr (muscle)	Cr (water)	BCF	Pb (muscle)	Pb (water)	BCF
AW1	0.024	0.041	0.585366	0.238	0.011	2.45952	1.953	2.044	1.259355
AW2	0.024	0.025	0.96	0.238	0.098	4.033613	1.953	1.925	2.065342
AW3	0.024	0.033	0.727273	0.238	0.108	3.055768	1.953	1.932	1.564653
AW4	0.024	0.037	0.648649	0.238	0.137	2.725414	1.953	1.929	1.395502
AW5	0.024	0.039	0.615385	0.238	0.033	2.58565	1.953	1.963	1.323937
BW1	0.019	0.028	0.678571	0.229	0.067	2.963194	1.873	1.931	1.582058
BW2	0.019	0.024	0.791667	0.229	0.092	3.45706	1.873	1.993	1.845734
BW3	0.019	0.025	0.76	0.229	0.052	3.318777	1.873	1.982	1.771905
BW4	0.019	0.023	0.826087	0.229	0.006	3.607367	1.873	1.975	1.925983
BW5	0.019	0.021	0.904762	0.229	0.08	3.950925	1.873	1.984	2.10941
CW1	0.021	0.02	1.05	0.212	0.021	4.95283	1.881	1.543	2.633084
CW2	0.021	0.021	1	0.212	0.022	4.716981	1.881	1.53	2.507699
CW3	0.021	0.031	0.677419	0.212	0.019	3.195374	1.881	1.511	1.698764
CW4	0.021	0.03	0.7	0.212	0.017	3.301887	1.881	1.611	1.755389
CW5	0.021	0.029	0.724138	0.212	0.023	3.415745	1.881	1.672	1.81592
Mean	0.021		0.776621	0.226333	0.0524	3.44934	1.902333	1.835	1.816982
SD	0.002		0.143203	0.011159	0.041626	0.71867	0.037238	0.197124	0.394005

Table 10.6 Bioconcentration (mg/kg) of heavy metals in the muscle of *Mystus gulio*

Water sample	Cd (muscle	Cd (water)	BCF	Cr (muscle	Cr (water)	BCF	Pb (muscle)	Pb (water)	BCF
AW1	0.015	0.041	0.365854	0.185	0.011	1.977587	1.955	2.044	1.011554
AW2	0.015	0.025	0.6	0.185	0.098	3.243243	1.955	1.925	1.658948
AW3	0.015	0.033	0.454545	0.185	0.108	2.457002	1.955	1.932	1.256779
AW4	0.015	0.037	0.405405	0.185	0.137	2.191381	1.955	1.929	1.120911
AW5	0.015	0.039	0.384615	0.185	0.033	2.079002	1.955	1.963	1.063428
BW1	0.013	0.028	0.464286	0.174	0.067	2.668309	1.872	1.931	1.425379
BW2	0.013	0.024	0.541667	0.174	0.092	3.113027	1.872	1.993	1.662942
BW3	0.013	0.025	0.52	0.174	0.052	2.988506	1.872	1.982	1.596424
BW4	0.013	0.023	0.565217	0.174	0.006	3.248376	1.872	1.975	1.735243
BW5	0.013	0.021	0.619048	0.174	0.08	3.557745	1.872	1.984	1.900505
CW1	0.016	0.02	0.8	0.179	0.021	4.469274	1.743	1.543	2.564127
CW2	0.016	0.021	0.761905	0.179	0.022	4.256451	1.743	1.53	2.442026
CW3	0.016	0.031	0.516129	0.179	0.019	2.883402	1.743	1.511	1.654276
CW4	0.016	0.03	0.533333	0.179	0.017	2.979516	1.743	1.611	1.709418
CW5	0.016	0.029	0.551724	0.179	0.023	3.082258	1.743	1.672	1.768364
Mean	0.015		0.538915	0.179333	0.0524	3.013005	1.856667	1.835	1.638021
SD	0.001		0.12366	0.004655	0.041626	0.714784	0.090287	0.197124	0.446268

A similar trend was observed for the bioconcentration factor. Chromium has the highest BCF, followed by chromium and lead. The average BCF values for lead, chromium, and cadmium are 1.638021, 3.013005, and 0.538915 L/kg, respectively. During the dry season, the bioconcentration factor is greater. The BCF (Cr) > BCF (Pb) > BCF (Cd) ranking indicates a decreasing order of the metal bioconcentration factor. This arrangement corresponds to the sequence in which heavy metals accumulate in fish muscle (Table 10.7).

This study analyzed the monthly concentrations of cadmium, chromium, and lead in the muscle of *Sardinella fimbriata* collected from the water. The results revealed that lead was the most accumulated metal in the fish during the entire sampling period, whereas cadmium remained the least accumulated metal. During the dry season, cadmium and chromium accumulate the most. Chromium had the highest bioconcentration factor (BCF), followed by chromium and lead. The average BCF values for lead, chromium, and cadmium are 2.215201, 3.986802, and 0.625662 L/kg, respectively.

Major contaminants include lead, cadmium, and chromium because of their toxicity, persistence, and propensity to bioaccumulate in the food chain (Kouamenan et al. 2020; Asim and Nageswara Rao 2021). They pose a threat to both people and aquatic ecosystems. These metals are most frequently evaluated in fish farming. A previous study revealed that heavy metals are present in water, but lead concentrations are higher than the highest levels permitted in water (Abdel-Baki et al. 2011; Alvarado et al. 2021). This may be because lead accumulates mostly in bottom sediments in surface waters, where lead concentrations are higher than those in water (Pintaeva et al. 2011; El-Korashy et al. 2018; Kumar et al. 2020; Li et al. 2017).

This study also revealed that sediments have far higher concentrations of chromium than water does, and water has higher concentrations of lead and cadmium. The propensity of metals to accumulate in sediments and biomagnify along aquatic food chains may account for this discrepancy (Moiseenko et al. 2020; Kouakou 2017). Fish consumption is crucial for evaluating heavy metal pollution in water (Coulibaly et al. 2012; Abarghoei et al. 2016; Kouamenan et al. 2020; Membere & Abdulwasiu 2020), as fish are vectors of heavy metal contamination in people (Hashim et al. 2014; Svobodová 1993). The muscle of these samples contains heavy metals, with lead being the most accumulated metal in this organ (Qadir and Malik 2011; Yasmeen et al. 2016; Mensoor and Said 2018).

Bioaccumulation factor (BSAF) values less than 1 indicate that the metals under study have essentially negligible bioaccumulation in fish (Nasir et al. 2020; Salati and Moore 2010). This might be related to the metal concentration in the sediments (Mutlu and Kurnaz 2018; Xiao et al. 2019; Zeng et al. 2020). Sediments are below the level at which organisms can accumulate metals in their bodies (Szefer et al. 1999; Usero et al. 2005; Zhao et al. 2012; Zhang et al. 2017; Zhang et al. 2021; Sanou et al. 2021), and these BSAF values <1 could explain the strong negative and substantial correlations between the concentrations of these metals in fish muscle and the sediments, suggesting that the concentrations in the sediment had no effect on bioaccumulation (Yi et al. 2011; Kouakou et al. 2016).

Table 10.7 Bioconcentration (mg/kg) of heavy metals in the muscle of *Sardinella fimbriata*

Water sample	Cd (muscle	Cd (water)	BCF	Cr (muscle)	Cr (water)	BCF	Pb (muscle)	Pb (water)	BCF
AW1	0.014	0.041	0.34146	0.143	0.011	2.387856	1.921	2.044	1.243028
AW2	0.014	0.025	0.56	0.143	0.098	3.916084	1.921	1.925	2.038565
AW3	0.014	0.033	0.42424	0.143	0.108	2.96673	1.921	1.932	1.544368
AW4	0.014	0.037	0.37837	0.143	0.137	2.646003	1.921	1.929	1.377409
AW5	0.014	0.039	0.35897	0.143	0.033	2.51031	1.921	1.963	1.306773
BW1	0.019	0.028	0.67857	0.154	0.067	4.406308	1.781	1.931	2.474064
BW2	0.019	0.024	0.79166	0.154	0.092	5.140693	1.781	1.993	2.886408
BW3	0.019	0.025	0.76	0.154	0.052	4.935065	1.781	1.982	2.770952
BW4	0.019	0.023	0.82608	0.154	0.006	5.364201	1.781	1.975	3.011904
BW5	0.019	0.021	0.90476	0.154	0.08	5.875077	1.781	1.984	3.298752
CW1	0.017	0.02	0.85	0.171	0.021	4.97076	1.743	1.543	2.851842
CW2	0.017	0.021	0.80952	0.171	0.022	4.734057	1.743	1.53	2.71604
CW3	0.017	0.031	0.54838	0.171	0.019	3.206942	1.743	1.511	1.839898
CW4	0.017	0.03	0.56666	0.171	0.017	3.31384	1.743	1.611	1.901228
CW5	0.017	0.029	0.58620	0.171	0.023	3.428111	1.743	1.672	1.966787
Mean	0.016	0.284	0.62566	0.156	0.0524	3.986802	1.815	1.835	2.215201
SD	0.002	0.0066961	0.19263	0.011922	0.041626	1.144838	0.079228	0.197124	0.683791

A substantial positive connection ($r = 0.81$) was found between fish muscle and cadmium in sediments, indicating that the tilapia body accumulates more of this metal in its flesh concentration in the sediments.

10.9 Conclusion

According to this investigation, there were significant quantities of lead and cadmium in the sediments and high concentrations in the fish farm water. Fish muscle typically has metal concentrations below predetermined thresholds because they are safe for ingestion by humans and comply with international law (FAO, WHO, and EU). Lead levels, however, were above these thresholds, posing a risk to consumers' health. The metals moved from the water to the fish, and the correlation analysis revealed that the amount of mercury in the water had a significant effect on the bioconcentration. For the metals under study, the bioaccumulation factor (BSAF) was nearly negligible, with the exception of the bioaccumulation of chromium metal in the laboratory. Nonetheless, the Pearson coefficient values demonstrated that the amount of Cd in the sediments affected its bioaccumulation in the muscle of the fish. The bioconcentration factor (BCF) results revealed that the transfer of metals from water to fish was greater than the transfer of metals from sediment to fish. This makes it possible to determine that water plays a major role in metal buildup in fish muscle. These findings show that fish have a high level of lead contamination and that fish have the capacity to both bioconcentrate and bioaccumulate, all of which pose concerns for fish survival and public health.

References

Abarghoei S, Hedayati A, Ghorbani R, Miandareh HK, Bagheri T (2016) Histopathological effects of waterborne silver nanoparticles and silver salt on the gills and liver of goldfish Carassius auratus. Int J Environ Sci Technol 13:1753–1760

Abdel-Baki AS, Dkhil MA, Al-Quraishy S (2011) Bioaccumulation of some heavy metals in tilapia fish relevant to their concentration in water and sediment of Wadi Hanifah, Saudi Arabia. Afr J Biotechnol 10(13):2541–2547

Ahuja S (2009) Handbook of water purity and quality. Academic Press, New York, pp 1–3

Ahmad W, Alharthy RD, Zubair M et al (2021) Toxic and heavy metals contamination assessment in soil and water to evaluate human health risk. Sci Rep 11:17006. https://doi.org/10.1038/s41598-021-94616-4

Ali MM, Ali ML, Islam MS, Rahman MZ (2016) Preliminary assessment of heavy metals in water and sediment of Karnaphuli River, Bangladesh. Environ Nanotechnol Mon Manag 5:27–35

Alvarado C, Cortez-Valladolid DM, Herrera-López EJ, Godínez X, Ramírez JM (2021) Metal bioaccumulation by carp and catfish cultured in lake chapala, and weekly intake assessment. Appl Sci 11(13):6087

Al-Weher SM (2008) Levels of heavy metal Cd, Cu and Zn in three fish species collected from the Northern Jordan Valley, Jordan. Jordan J Biol Sci 1(1):41–46

Asim M, Nageswara Rao K (2021) Assessment of heavy metal pollution in Yamuna River, Delhi-NCR, using heavy metal pollution index and GIS. Environ Monit Assess 193(2):103–107

Balali-Mood M, Naseri K, Tahergorabi Z, Khazdair MR, Sadeghi M (2021) Toxic mechanisms of five heavy metals: mercury, lead, chromium, cadmium, and arsenic. Front Pharmacol 227

Begum A, Ramaiah M, Harikrishna, Khan I, Veena K (2009) Heavy metal pollution and chemical profile of Cauvery River water. J Chem 6(1):47–52

Bhat NA, Ghosh P, Ahmed W, Naaz F, Darshinee AP (2023) Heavy metal contamination in soils and stream water in Tungabhadra basin, Karnataka: environmental and health risk assessment. Int J Environ Sci Technol 20(3):3071–3084

Coulibaly S, Atse BC, Koffi KM, Sylla S, Konan KJ, Kouassi NGJ (2012) Seasonal accumulations of some heavy metal in water, sediment and tissues of black-chinned tilapia Sarotherodon melanotheron from Biétri Bay in Ebrié Lagoon, Ivory Coast. Bull Environ Contam Toxicol 88:571–576

El-Korashy SA, Ali NA, Al-Kenawy DA, Mousa IM (2018) Some heavy metals concentrations in water and tilapia (Oreochromis niloticus) muscles of fish farms supplied by three different water sources

Fong PP, Molnar N (2008) Norfluoxetine induces spawning and parturition in estuarine and freshwater bivalves. Bull Environ Contam Toxicol 81:535–538

Gabrielyan AV, Shahnazaryan GA, Minasyan SH (2018) Distribution and identification of sources of heavy metals in the Voghji River basin impacted by mining activities (Armenia). J Chem

Ghorade IB, Lamture SV, Patil SS (2014) Assessment of heavy metal content in Godavari river water. Int J Res Appl Nat Soc Sci 2(6):23–26

Giri S, Singh AK (2014) Assessment of surface water quality using heavy metal pollution index in Subarnarekha River, India. Water Qual Expo Health 5:173–182. https://doi.org/10.1007/s12403-013-0106-2

Gowd SS, Reddy MR, Govil PK (2010) Assessment of heavy metal contamination in soils at Jajmau (Kanpur) and Unnao industrial areas of the ganga plain, Uttar Pradesh, India. J Hazard Mater 174(1–3):113–121

Gupta A, Rai DK, Pandey RS, Sharma B (2009) Analysis of some heavy metals in the riverine water, sediments and fish from river Ganges at Allahabad. Environ Monit Assess 157:449–458

Hashim R, Song TH, Muslim NZM, Yen TP (2014) Determination of heavy metal levels in fishes from the lower reach of the Kelantan River, Kelantan, Malaysia. Trop Life Sci Res 25(2):21–28

Kanmony C (2009) Human rights and health care. Mittal Publication, New Delhi, pp 73–76

Kennish JM, Sharp-Dahl JL, Chambers KA, Thrower F, Rice SD (1992) The effect of a herring diet on lipid composition, fatty acid composition, and cholesterol levels in the muscle tissue of pen-reared Chinook salmon (Oncorhynchus tshawytscha). Aquaculture 108(3–4):309–322

Khan MA, Ghouri AM (2011) Environmental pollution: its effects on life and its remedies. Researcher World J Arts Sci Comm 2(2):276–285

Kouamenan NZM, Coulibaly S, Atse BC, Goore BG (2020) Bioaccumulation of some heavy metals in the tissues of two species of cichlidae (Hemichromis fasciatus and hybrid Tilapia zilliix Tilapia guineensis) collected from the western part of the Ebrie lagoon, Côte d'Ivoire

Kouakou AS, Coulibaly SL, Coulibaly L, Savane I (2016) Evolution Temporelle del'utilisation des pesticides en agriculture tropicale dans le bassin versant de la Marahoué, Côte d'Ivoire. Int J Innov App Stud 14(1):121–131. URL: http://www.ijias.issr-journals.org

Kouakou AR (2017) Evaluation de la spéciation, de la biodisponibilité et de la toxicité potentielle des métaux cadmium, cuivre, plomb et zinc dans les sédiments du canal de Vridi (Côte d'Ivoire). Thèse de Doctorat, Université Félix Houphouët Boigny, 180 p

Kumar A, Dua A (2009) Water quality index for assessment of water quality of river Ravi at Madhopur (India). Glob J Environ Sci 8(1). https://doi.org/10.4314/gjes.v8i1.50824

Kumar V, Sharma A, Kumar R, Bhardwaj R, Kumar Thukral A, Rodrigo-Comino J (2020) Assessment of heavy-metal pollution in three different Indian water bodies by combination of multivariate analysis and water pollution indices. Hum Ecol Risk Assess Int J 26(1):1–16

Li H, Lin L, Ye S, Li H, Fan J (2017) Assessment of nutrient and heavy metal contamination in the seawater and sediment of Yalujiang Estuary. Mar Pollut Bull 117(1–2):499–506

Membere E, Abdulwasiu M (2020) Heavy metals concentration in water, sediment, and fish around Escravos River, Nigeria. World J Res Rev 10:28–37

Mensoor M, Said A (2018) Determination of heavy metals in freshwater fishes of the Tigris River in Baghdad. Aust Fish 3(2):23
Mico C, Peris M, Sanchez J, Recatala L (2006) Heavy metal content of agricultural soils in a Mediterranean Semiarid area: the Segura River Valley, Alicante, Spain. Span J Agric Res 4:363–372
Moiseenko TI, Gashkina NA, Dinu MI (2020) Metal bioavailability in northern low-salinity water: case study of lakes in the Kola region, Russia. Environ Res Lett 15(9):095005
Mutlu E, Kurnaz A (2018) Assessment of physicochemical parameters and heavy metal pollution in Çeltek Pond water. Indian J Geo-Mar Sci 47(6):1185–1192
Nasir M, Iqbar I, Munira M, Muchlisin ZA, Saiful S, Suhendrayatna S, Erdiwansyah E (2020) Investigation of heavy metals in river water, sediments, and fish in Krueng Geumpang, Pidie regency, Aceh Province. J Adv Res Fluid Mech Therm Sci 75(1):81–93
Papatilippaki AK, Kotti ME, Stavroulakis GG (2008) Seasonal variations in dissolved heavy metals in the Keritis River, Chania, Greece. Global Nest Int J 10(3):320–325
Pintaeva ET, Bazarsadueva SV, Radnaeva LD, Petrov EA, Smirnova OG (2011) Content and character of metal accumulation in fish of the Kichera River (a tributary of Lake Baikal). Contemp Probl Ecol 4:64–68
Prasad S, Saluja R, Joshi V, Garg JK (2020) Heavy metal pollution in surface water of the upper Ganga River, India: human health risk assessment. Environ Monit Assess 192(11):742
Proshad R, Islam S, Tusher TR, Zhang D, Khadka S, Gao J, Kundu S (2021) Appraisal of heavy metal toxicity in surface water with human health risk by a novel approach: a study on an urban river in vicinity to industrial areas of Bangladesh. Toxin Rev 40(4):803–819
Qadir A, Malik RN (2011) Heavy metals in eight edible fish species from two polluted tributaries (Aik and Palkhu) of the River Chenab, Pakistan. Biol Trace Elem Res 143:1524–1540
Qishlaqi A, Moore F (2007) Statistical analysis of accumulation and sources of heavy metals occurrence in agricultural soils of Khoshk River Banks, Shiraz, Iran. Am Eurasian J Agric Environ Sci 2(5):565–573
Salati S, Moore F (2010) Assessment of heavy metal concentration in the Khoshk River water and sediment, Shiraz, Southwest Iran. Environ Monit Assess 164:677–689
Sanou A, Coulibaly S, Coulibaly M, N'Goran N'dri S, Célestin Atse B (2021) Assessment of heavy metal contamination of fish from a fish farm y bioconcentration and bioaccumulation factors. Egypt J Aquat Biol Fish 25(1):821–841
Sarkar B (2005) Heavy metals in the environment. Taylor and Francis, New York, pp 33–41
Simeonov V, Simeonova P, Tsakovski S, Lovchinov V (2010) Lake water monitoring data assessment by multivariate statistics. J Water Resour Prot 2(4):353
Simpson S, Batley G (eds) (2016) Sediment quality assessment: a practical guide. CSIRO Publishing, Clayton
Svobodová Z (1993) Water quality and fish health (No. 54). Food & Agriculture Org
Szefer P, Ali AA, Ba-Haroon AA, Rajeh AA, Gełdon J, Nabrzyski M (1999) Distribution and relationships of selected trace metals in molluscs and associated sediments from the Gulf of Aden, Yemen. Environ Pollut 106(3):299–314
Siziba N, Mwedzi T, Muisa N (2021) Assessment of nutrient enrichment and heavy metal pollution of headwater streams of Bulawayo, Zimbabwe. Phys Chem Earth Parts A/B/C 122:102912
Usero J, Morillo J, Gracia I (2005) Heavy metal concentrations in molluscs from the Atlantic coast of southern Spain. Chemosphere 59(8):1175–1181
Voet D, Voet G, Pratt C (2008) Principles of biochemistry. Wiley, New York
Xiao H, Shahab A, Li J, Xi B, Sun X, He H, Yu G (2019) Distribution, ecological risk assessment and source identification of heavy metals in surface sediments of Huixian karst wetland, China. Ecotoxicol Environ Saf 185:109700
Yasmeen K, Mirza MA, Khan NA, Kausar N, Rehman AU, Hanif M (2016) Trace metals health risk appraisal in fish species of Arabian Sea. Springerplus 5:1–7
Yi Y, Yang Z, Zhang S (2011) Ecological risk assessment of heavy metals in sediment and human health risk assessment of heavy metals in fishes in the middle and lower reaches of the Yangtze River basin. Environ Pollut 159(10):2575–2585

Zeng J, Han G, Yang K (2020) Assessment and sources of heavy metals in suspended particulate matter in a tropical catchment, Northeast Thailand. J Clean Prod 265:121898

Zhang C, Shan B, Tang W, Dong L, Zhang W, Pei Y (2017) Heavy metal concentrations and speciation in riverine sediments and the risks posed in three urban belts in the Haihe Basin. Ecotoxicol Environ Saf 139:263–271

Zhang H, Liu S, Wu K, Cui J, Zhu A, Zhang Y et al (2021) Distribution and assessment of heavy metal contents in surface sediments of the western Sunda shelf. Mar Pollut Bull 168:112433

Zhao S, Feng C, Quan W, Chen X, Niu J, Shen Z (2012) Role of living environments in the accumulation characteristics of heavy metals in fishes and crabs in the Yangtze River Estuary, China. Mar Pollut Bull 64(6):1163–1171

Interstate Dispute on Mahanadi Water: Political Conflicts and Eco-ethical Solutions

11

Narayan Jena and Krishna Pratap Singh

Abstract

The Mahanadi water dispute between Chhattisgarh and Odisha pushed the latter into a state of severe water scarcity. Chhattisgarh's noncompliance with riparian rights and territorialized monopolistic approach in the management and governance of the Mahanadi water resource has put the soil and river ecology under threat. Despite the call for judicial intervention and a constitutional role, no viable solution has been brought about to end the asymmetric access to the Mahanadi water between the upstream and downstream states. Although the scant flow of water in the downstream state of Odisha clearly poses a threat to the land–water–food–life nexus, which is undoubtedly detrimental to the national economy, the flow of water is determined by political power. Water disputes have become part of the political agenda in both states. As a result, the ecological side of the river is largely neglected and less discussed, yet claiming the right over the river has become the dominant subject matter for retaining the pride of the state and the ruling political party. As a result, the conflict is perpetuated. This present paper intends to foreground and examine the factors responsible for interstate water disputes and will explain how eco-ethical solutions can address political issues related to Mahanadi water and mitigate river and soil ecology crises.

Keywords

Mahanadi · Water disputes · Eco-ethics · Political conflicts

N. Jena
Faculty of Arts, Communication and Indic Studies, Sri Sri University, Govindpur, Odisha, India

K. P. Singh (✉)
Institute of Indian and Foreign Languages, MGM University, Chhatrapati Sambhajinagar, Maharashtra, India

J. P. Bhatt, A. K. Mishra (eds.), *Mahanadi River*,
https://doi.org/10.1007/978-981-96-1048-8_11

11.1 Introduction

Water, the life-sustaining force of nature that freely flows in rivers and lakes, has paramount economic value. The International Conference on Water and the Environment in Dublin, in January 1992, upheld a principle that regards water as having "economic value." It is recommended that every individual has basic rights to have access to "clean water." Therefore, the proper management of water for achieving efficient and equitable utilization is fundamental for the protection of water resources. Ironically, however, the economic value of water places it at the center of many political debates (McGee 1909) because of the steep surge in the demand for water for the promotion of economic development. The rise in the demand for water has created a situation of conflict and competition over water resources among different bodies to grab the maximum by using the instrument of power. A river with its complete capacity fortifies various economic activities and regional developments by fostering agricultural, commercial, and industrial growth (Pandya and Sharma 2023; Wang et al. 2023). A healthy river not only contributes to the present economic infrastructure but also ensures the security of future generations against the backdrop of increasing water scarcity and climate change. It contributes to the ecological health of soil, ocean, river basins, etc. (Ren et al. 2023). The dynamic ecosystem of rivers, through the supply of essential nutrients, organic substances, and minerals, supports aquatic richness, soil fertility, and terrestrial ecosystem balance. Slight changes in the functioning of river ecosystems adversely affect the physical environment and the taxonomic and functional diversity of aquatic life (Stubbington et al. 2024). Any ecological alteration due to the regulation of the downstream water flow of rivers leads to detrimental changes and disruptions in riverine, terrestrial, and ocean life (Durrant et al. 2023; Jacquet et al. 2016). Both natural biodiversity and the human community are highly dependent on the well-being of river systems. The socioeconomic stability of many communities and the cultural dynamics of different regions profoundly draw from rich river systems. Daniel et al. (Daniell and Moggridge 2024) find the river as the lifeline for cultural identity, and Jackson et al. (Jackson and Barber 2013) highlight ecology-based identity as the ground for cultural recognition and maturity.

11.2 Geophysical and Cultural Significance of the Mahanadi River

The Mahanadi River, the longest river of Odisha (a state in the eastern part of India), originates from the mountain terrain of Chhattisgarh and runs 560 miles (approximately 900 kilometers) in both states before flowing into the Bay of Bengal. Geophysically present in the state of Chhattisgarh and Odisha, it supports many ecosystems, from plateaus to planes and from agricultural to aquatic biodiversity. The river basin of Mahanadi, characterized by its stretch of alluvial soil, supports the cultivation of paddy, wheat, mazes, varieties of millet, and cash crops. The vast delta regions of Mahanadi provide rich habitats for different exotic animals and plants, which highlights the ecological significance of the river. It maintains hydrological balance in the surrounding landscapes and freshwater habitat; it supports the health of mangroves and river basin soil; and it also maintains balance in underground aquifers and water bodies. The vast stretch of the river's drainage area, which is approximately 51,000 square miles (132,100 square kilometers), enriches the riverbank with fine nutrients, which are helpful for soil health. These findings indicate that the system of Mahanadi is not a simple ecological system; rather, it is intricately connected to many other complex biological communities, including those of human ecology. Therefore, in addition to preserving riparian biodiversity, rivers contribute to regional culture and ecological sustainability.

It is widely acknowledged by the people of Odisha that the river significantly contributes to the constitution of people's regional identities, which is exemplified by local festivals, rituals, and religious performances associated with the river. Many local narratives, legends, and folktales, which revolve around Mahanadi, highlight the importance of the river in the cultural lives of Odisha. The complex relationship between Mahanadi and Odisha's heritage sites is evident in local practices, which hinge upon the seasonal cycle of the river. It is evident that for the people of Odisha, Mahanadi is not a mere source of water but a natural heritage that has spiritual and cultural significance. Thus, it reserves a significant place in the collective consciousness or cultural memory of the Odishan people. However, massive urbanization and industrialization with infrastructural development threaten the integrity of the river. Therefore, there is a need for systematic management of the river for its sustainability and for controlling the exhaustive use of the river by the industries promoted by the states.

11.3 Impacts of Dams and Hydroprojects on the River Dynamics of Mahanadi

With the recognition of the commercial potential of the Mahanadi River by Odisha, Madhya Pradesh, and Chhattisgarh for hydroprojects, irrigation, and industrial supply, many major dams, such as Hirakud, Ravishankar Sagar, Dudhawa, Sondur, Hasdeo Bango, and Tandula, were built in the upper and lower riparian states along the river. By the end of 2015, as many as 254 major dams and thousands of barrages and irrigation projects had been built on Mahanadi by the two states. Among the total major projects, 168 are situated in the upper riparian state of Chhattisgarh, and Odisha managed to build 86 projects. Therefore, some sections of society believe that Odisha has failed to fully utilize river water because of the lower number of projects. In addition, as per the declaration of the government of Odisha, seven new barrages were constructed across Mahanadi, downstream of the Hirakud reservoir.

Dams were constructed for the purpose of flood control, irrigation, and drinking water supply to cities, but they have also supplied water to the industries fostered by the states. The barrages specifically designed for industrial water supply place pressure on the natural water capacity of the river, which consequently decreases its capacity to a minimum. Water extractions above the permissible level adversely affect the river's ecological dynamics. In addition, these extractions disturb the natural habitats of many living organisms and aquatic animals and displace many local communities. The maximum use of river water in the upper riparian state leaves the river in a dry and dead state in the lower stream regions. As it is against the riparian rights of the downstream state of Odisha, many political debates have occurred between the two states, and the river Mahanadi has become the center of many political conflicts and legal battles. While playing the political games of riparian rights and the economic gamble of profit and loss, both the states and their political and legal institutions become myopic about habitat fragmentation, the dynamics of aquatic species, and the alteration of the biotic systems of the river.

11.4 Overview of Disputes Between Odisha and Chhattisgarh Over Mahanadi Water

The dispute between Odisha and Chhattisgarh over Mahanadi began with the claim of Odisha that the illegal construction of several barrages across Mahanadi by the upper riparian state of Chhattisgarh seriously affects the inflow of water into the Hirakud reservoir in Odisha during the non-monsoon season. These constructions not only are "illegal" but also kill the rich ecosystem of the Mahanadi River. As the debate surged, the Supreme Court directed the union government to form a river dispute tribunal to bring about a resolution pertaining to the fair and equitable distribution of Mahanadi water. Odisha's allegation of the scant flow of water to the Hirakud reservoir from Chhattisgarh triggered the dispute. The Hirakud dam is one of the important generators of hydropower in Odisha, a large supplier of water for industrial and urban demands and has the objective of ensuring flow in Mahanadi to meet drinking water, agricultural, and ecological needs. When the run-off at the Hirakud dam has substantially dwindled, the plans of both Odisha and Chhattisgarh to maximize the utilization of Mahanadi River water appear to be ecologically damaging and politically conflicting. The promotion of massive industrialization in both states by tapping the rich mineral resources of the area redoubled the use of water.

Odisha's utilization plan, which was submitted at the tripartite meeting of September 2016, shows that the plan is much greater than the current average runoff it receives at the site. Similarly, Chhattisgarh has atrociously planned to utilize an amount of water that it is unlikely to obtain in a normal year. The Odisha government has quoted CWC sources to further allege that the "planned utilization" of Chhattisgarh through "ongoing" and "future" projects is much greater than the average normal potential of the Mahanadi River in the whole state. Such "plans" are the testimonies of irrational and irresponsible behavior of both the Odisha and Chhattisgarh governments toward the utilization of river water. Again, both states have plans to extract maximum river water to cater to the needs of industries, which leads to legal and political conflicts between the states in terms of the reduced flow of water. Conflicts over Mahanadi water since 1953 have centered around the demand of both states to maximize river water exploitation. Thus, despite a series of discussions and the formation of different tribunals, no concrete solutions exist to address this problem. Even the legal battle between the two states has not resulted in any solution to the issue. Although legal battle can promise riparian rights to the state, it never recognizes the rights of the river to be alive in a healthy state. Thus, in view of the inability of the political and legal provisions, an ethical consideration may be viewed as a modality to find a solution to end the conflict and save the Mahanadi ecosystem.

11.5 Ethical Approaches to the Use of Mahanadi Water

Ethics concerning water use denotes the underlying principles that influence the responsible behavior of citizens toward the sensible use and conservation of water resources. This set of normative values is applied to understand the egregious human activities that cause irreparable damage to water ecology. The greed of both Odisha and Chhattisgarh promotes developmental activities that create an ecological mess on a large scale. Although ethics may not necessarily provide direct solutions to problems, they establish an enabling context to unleash creative and thoughtful solutions to problems. As Simpson (2017) noted, ethics can stimulate innovation and awareness, which can then be used for better decisions on water management and governance. As an instrument for approaching decisions about water policies and water use, it instills a sense of responsibility to protect rivers from human-induced damage without compromising people's property and lives (Groenfeldt 2017). As a manifestation of individual values that guide humans in moral action, water ethics makes people think of the health of water as everybody's concern and responsibility because it is a valuable component of the environment. It also highlights the "trajectories of ecological degradation" that are caused by political institutions, ideologies, and practices (Groenfeldt 2019, p. 14). Furthermore, Groenfeldt suggested that water ethics combines "water integrity," "water stewardship," and "water justice" (pp. 13–15). When integrity involves the idea of respecting everyone's right to water and viewing equitable access to water resources, stewardship refers to the ethical and moral responsibility of humans to use water bodies in such a way that they are not corrupt. Finally, water justice views the rights of all, including the underprivileged and nonhumans. It also views nature; moreover, natural resources have their right to exist the way they naturally are. Thus, there should not be any human intervention that defiles and alters their course of functioning.

In line with this view, Mahanadi may argue that Mahanadi has its own rights to exist and flow; as a result, others' rights to water resources will be protected. It is imperative to develop new ways to value the water ecosystem of Mahanadi. The existing systems, which are politically governed, view the water resources of rivers as commodities with potential for industrial growth. Thus, the political systems of the states legitimize their rights to maximize the exploitation of river water to support their infrastructure. The ethically responsible behavior of both states in using Mahanadi water will certainly limit the overexploitation of the river's water with respect to the rights of nature (Ballet and Bazin 2017). As dams and barrages are responsible for altering river ecosystems but promote economic growth, they should be designed in such a way that the restriction of river flow is minimal. Economic activities will continue by allowing rivers to flow and allowing aquatic life to flourish. Importantly, infrastructure designed at the cost of ecological health is never sustainable (Groenfeldt 2019, pp. 22–23). Notably, Mahanadi is important not only because its water can be used for industrial purposes but also because of its ecological and cultural value. Therefore, ethics-based water governance is needed to support the equitable distribution of Mahanadi water between the upper and lower riparian states because perennial flow is an ecological necessity for the sustenance of river, soil, and ocean ecology. It is not eco-ethically justified to mathematically calculate the use of water by state based on the catchment areas in their respective states. The water distribution logic or politics between states should not affect river flow; rather, river ecology in the lower riparian state needs to be respected and conserved (p. 22). Therefore, Mahanadi, as an organic system, should not be completely engineered; its natural functions need to be retained undisturbed, as they are necessary for sustainability. The ecological values are expected to be instilled in the political system to create good water governance and Mahanadi River conservation, which will protect the sociocultural life of local people and ecosystems. When river exploitation in the upper riparian state of Chhattisgarh leaves the river parched, the maximum use of water by the lower riparian state of Odisha to meet its urban and agricultural needs will completely manipulate the "hydrological cycle of nature" (Blackbourn 2006).

Blackbourn (2006) suggested the process of "renaturalization" of rivers, which involves the implementation of the best possible methods to preserve or restore the sanctity of the ecological system. In the context of Mahanadi, the river management system needs to be ethically and morally governed to maintain the "good ecological status" of the river and to keep the river flowing without being completely gridlocked. John Wesley Powell, in his study on the Colorado River, suggested the use of scientific methods for the protection of river flow. However, conservationist John Muir viewed recognition of the sanctity of nature as a method for protecting nature (Muir 1912). Reisner (1993) emphasized legal provisions for "leaving river intact." However, both science and politics need to be governed by a set of ethical principles that are oriented toward ecology. As Reisner and Bates (1990) noted, the bleak picture of the environment needs to be altered. The alteration of the picture can be made possible through the introduction of new jurisprudence, new scientific knowledge, and new epistemes to view sustainability over industrial and infrastructural development. Under the new episteme, Mahanadi will be viewed as a living system, not an object to be controlled and conquered through state power. This new jurisprudence will lead to a resurgence in the construction of large dams as an ecological crime.

11.6 Water Cooperation as an Ethical Method to Resolve Water Conflicts

Owing to the scant flow of water in Mahanadi, conflict between the two states to obtain a larger chunk is deemed inevitable. However, in a federal structure such as India, the states are programmed to function in interdependence with each other for cultural, economic, and ecological sustenance. However, the conflicts between the states are mostly political in nature. Political conflicts adversely affect economic and ecological dynamism. The politics of Mahanadi water sharing has caused the river to be plundered. Chhattisgarh's geographical location as the upper riparian state creates the opportunity for the state to exploit the larger share of water without paying attention to the lower riparian river ecology and Odisha's riparian rights to have fair access to Mahanadi water resources. Such politics need to be governed by ethics, which gives respect to water resources and values river ecology and people's demand for healthy cultural life (Brown and Schmidt 2010; Kallhoff 2017). Therefore, water cooperation may be viewed as a method for resolving water conflicts in reasonable ways. Water cooperation, according to Kallhoff (2017), is a value action based on the argument that resources can be mutually shared among stakeholders to resolve water conflicts (p. 93). Here, both Chhattisgarh and Odisha need to reach an agreement that the basic ecological functionality of Mahanadi needs to be kept intact and thus should be valued above politics and economic activities.

The economic and developmental activities of Mahanadi need to be guided by the principles of fairness and equitability in the distribution of water resources between the states with respect to the cultural, ecological, and aesthetic values associated with the river. The Mahanadi water needs to be considered a "shared resource." As a result, the river should be protected by both states. Therefore, the maximum exploitation of water by Chhattisgarh or Odisha is ethically unacceptable because such action disturbs river ecology and aesthetics, which do not belong to any state. A scant supply of water downstream certainly restricts eco-functionality (Ripl 2003). The ecological functionality, as a matter of ethical concern, needs to be removed from the political and economic ambit because politics do not give value to the river's functionality. The ecological functionality of a river includes the rights of all living beings to obtain fair access to it (Gleick 1998). It also reflects the practices of various communities that have grown in connection with rivers.

11.7 Ethical Concerns on the Culture Associated with the Mahanadi

Water ethics focuses on respect for community life and culture. It views water as part of cultural practices (Gyawali 2009). In other words, cultural identities are widely associated with natural heritages. The Mahanadi River, historically and culturally, is connected to the lives of the Odia people. Therefore, the alteration of the river in any form indicates the disintegration of the cultural life of the people of Odias and some ethnic communities of Chhattisgarh. The aesthetic and spiritual values of rivers are based on cultural values. In addition, many communities have evolved symbiotic relationships with Mahanadi. Therefore, their sociocultural lives are inextricably connected to the river. This connection stems from the historical past and continues to be present in contemporary city life.

If cultural identity is "fundamental to human well-being" (Kellert 2012) and, as per the UN's declaration of 2007, indigenous people's cultural well-being needs to be preserved, Mahanadi, as a natural system that historically determines cultural identity, needs to be protected from all forms of alteration. The river needs to be left alone to become ecologically functional. The idea of controlling a river to make it useful through the imposition of engineering structures and modifications of river streams needs to be reassessed to ensure river flow. It is important to extend moral consideration and ethical rights to water itself. The water of the river needs to be used for legitimate purposes. It is unreasonable for any institution to control and accumulate it. Water cannot be left to function as a property of private bodies that will decide the course of its flow; rather, it should be left to flow the way it does. There is a need for ethical justification in the formulation of river policies for maintaining the health of water ecosystems.

11.8 Conclusion

The water conflicts between Odisha and Chhattisgarh require ethical imperatives to reach resolution. The dialogue for maintaining equitable distribution of water while respecting the rights of the river and local communities needs to be initiated. Shifting the focus from a developmental economy to an eco-ethical economy will lead to changes in policies, which will certainly protect the rights of all associated with the river. While making any decision pertaining to river water, the views of local people can be the best way to resolve the conflict. Above all, the ecological importance of rivers needs to be acknowledged because sustainable water ecology can create a sustainable economy and healthy social system. Protection of the river ecosystem should be the ethos of the political system. Otherwise, debate will occur on a dead river, and the political parties will play blame games over the death of the river.

References

Ballet J, Bazin D (2017) Hans jonas: bridging the gap between environmental justice and environmental ethics? Environ Ethics 39(2):175

Blackbourn D (2006) The conquest of nature: water, landscape, and the making of modern Germany. Norton, New York

Brown PG, Schmidt JJ (eds) (2010) Water ethics: foundational readings for students and professionals. Island Press, Washington

Daniell KA, Moggridge B (2024) Indigenous water engineering and aquaculture systems in Australia: the Budj Bim cultural landscape and Baiame's Ngunnhu (the Brewarrina aboriginal fish traps). Blue Papers 3(1)

Durrant E, Howson P, Puttick B, Potts S, Shennan-Farpón Y, Sari N et al (2023) Existing evidence on the use of participatory scenarios in ecological restoration: a systematic map. Environ Evid 12(1):27

Gleick PH (1998) The human right to water. Water Policy 1:487–503

Groenfeldt D (2017) Incorporating ethics into water decision making. In: Ziegler R, Groenfeldt D (eds) Global water ethics: towards a global ethics charter. Routledge, New York, pp 75–90

Groenfeldt D (2019) Water ethics: a value approach to solving the water crisis. Routledge, New York

Gyawali D (2009) Water and conflict: whose ethics is to prevail? In: Llamas MR, Martinez-Cortina L, Mukherji A (eds) Water ethics. Marcelino Botin water forum 2007. CRC Press, Boca Raton, pp 13–24

Jackson S, Barber M (2013) Recognition of indigenous water values in Australia's Northern Territory: current progress and ongoing challenges for social justice in water planning. Plan Theory Pract 14(4):435–454

Jacquet C, Moritz C, Morissette L, Legagneux P, Massol F, Archambault P, Gravel D (2016) No complexity–stability relationship in empirical ecosystems. Nat Commun 7(1):12573

Kallhoff A (2017) Transcending water conflicts: an ethics of water cooperation. In: Ziegler R, Groenfeldt D (eds) Global water ethics: towards a global ethics charter. Routledge, New York, pp 91–106

Kellert S (2012) Birthright: people and nature in the modern world. Yale University Press, New Haven

McGee WJ (1909) Water as a resource. Ann Am Acad Pol Soc Sci 33(3):37–50. https://doi.org/10.1177/000271620903300304

Muir J (1912) The Yosemite. Century, New York

Pandya AB, Sharma P (2023) River Basin development for water and food security. Irrigation Drainage 72(2):584

Reisner M (1993) Cadillac Desert: the American west and its disappearing water. Penguin Books, New York

Reisner M, Bates S (1990) Overtapped oasis: reform or revolution for Western water. Island Press, Washington

Ren S, Zhao H, Zhang H, Wang F, Yang H (2023) Influence of natural and social economic factors on landscape pattern indices – the case of the Yellow River Basin in Henan Province. Water 15(23):4174

Ripl W (2003) Water: the bloodstream of the biosphere. Philos Trans R Soc Lond Biol Sci 358:1921–1934

Simpson P (2017) A sense of the cycling environment: felt experiences of infrastructure and atmospheres. Environ Plann A Econ Space 49(2):426–447

Stubbington R, England J, Sarremejane R, Watts G, Wood PJ (2024) The effects of drought on biodiversity in UK river ecosystems: drying rivers in a wet country. Wiley Interdiscip Rev Water 11:e1745

Wang Q, Gao Y, Gong T, Liu T, Sui Z, Fan J, Wang Z (2023) Dam surface deformation monitoring and analysis based on PS-InSAR technology: a case study of xiaolangdi reservoir dam in China. Water 15(18):3298

Water Governance: An Overview of the Interstate Political Conflict Pertaining to the Mahanadi River

12

Vijaya Lakshmi Mohanty and Hitanshu Shekhar Mohanta

Abstract

The challenges pertaining to water governance of the Mahanadi River between the state governments of Odisha and Chhattisgarh date back to more than 8 years ago. This is a clear example of the intricate and politicized nature of interstate water conflicts in the nation. Rooted in the privatization and diversion of water for industrial purposes, the dispute has exacerbated water scarcity in Odisha, particularly affecting its agriculture and drinking water supply. Political maneuvering has further complicated resolution efforts. Thus, this paper examines the historical context, political and industrial influences, and socioenvironmental impacts of the dispute. It proposes a framework for resolving the conflict, emphasizing the need for cooperative federalism, equitable water allocation, and sustainable water governance within constitutional and ecological contexts.

The Mahanadi River holds great importance to the farmers and industries in Odisha and Chhattisgarh. Tensions between the two states arose when Chhattisgarh initiated plans to construct barrages along the river. Chhattisgarh's rationale for the barrages was to prevent excess floodwater from flowing unused into the sea without disrupting the natural flow of the river. However, the Odisha government strongly opposed the construction of these dams and barrages, citing the welfare of farmers in the state as a primary concern. While both governments claim that their actions are in the interest of farmers and agriculture, in reality, industries in both states consume the majority of their water resources.

V. L. Mohanty (✉)
Human Resource Development Centre (HRDC), Sri Sri University, Cuttack, Odisha, India
e-mail: vm1@srisriuniversity.edu.in

H. S. Mohanta
School and Mass Education Department, Government of Odisha, Bhubaneswar, Odisha, India

J. P. Bhatt, A. K. Mishra (eds.), *Mahanadi River*,
https://doi.org/10.1007/978-981-96-1048-8_12

Keywords

Water governance · Mahanadi River discharge · Mahanadi River Basin · Water management

12.1 Introduction

The Mahanadi River dispute between Chhattisgarh and Odisha illustrates the complex and politically charged nature of interstate water conflicts in India. Owing to the privatization and diversion of water resources for industrial use, this dispute has worsened water scarcity, particularly in the lower riparian region of Odisha. Political agendas and the ecological deterioration of the river basin further intensified the conflict. This paper examines the interplay of three Ps: privatization, politics, and power dynamics in the Mahanadi dispute, addresses the socioenvironmental effects on local communities, and offers a framework for conflict resolution within a constitutional and ecological framework.

The Mahanadi River, which flows through central-east Bharat, plays a pivotal role in sustaining both Chhattisgarh and Odisha. Burgeoning water governance issues between two neighbouring states have become major environmental and political issues. The objective is to examine the root cause of the conflict, the ensuing ecological and social challenges, and potential pathways toward a sustainable resolution. The focus on water governance highlights the complex interplay of legal, political, and environmental factors in managing shared water resources.

12.2 Review of the Literature

An array of literature on the topic was surveyed from various sources, including the government, to understand the veracity of the dispute from different perspectives. In his 2005 article, "Water Governance: A Historical Understanding of Mahanadi River Basin, Orissa," S.K. Mahapatra discusses the Hirakud Command Area within the Mahanadi River Basin. He uses examples from various districts in Odisha to reinforce his arguments. Mahapatra called for greater investment in water resources and better water management practices, arguing that such initiatives would not only ensure the sustainable use of water but also improve livelihoods in rural areas, especially economically disadvantaged areas.

The Mahanadi water conflict between two neighboring states intensified because of competing demands for water, particularly from industrial sectors. Odisha Government accuses Chhattisgarh Government of constructing upstream barrages and reducing the river's downstream flow, which affected their agriculture. Another article in *Economic and Political Weekly* emphasized the need for improved water governance, interstate cooperation, and a tribunal to resolve the conflict. It also highlights the environmental and socioeconomic impacts of the dispute on both states and calls for sustainable solutions (Swain 2017). Heinrich-Böll-Stiftung

(2018) focused on the environmental degradation and industrial pressures in the Mahanadi River Basin, exacerbated by coal mining and thermal power plants. This emphasized the need for sustainable water management practices in both Odisha and Chhattisgarh, highlighting the environmental impact of industrial activities. The Water Conflict Forum (2017) discusses the need for integrated water management practices in the Mahanadi River Basin. It addresses how policy interventions could improve water allocation between the agricultural and industrial sectors, with a focus on addressing the challenges faced by Odisha and Chhattisgarh.

K.C. Ratha (2019) explored the complexities of water governance in the Mahanadi River Basin, focusing on conflicts between two neighboring states, Odisha and Chhattisgarh, over the construction of dams. This highlights the challenges in balancing industrial water usage and irrigation needs, criticizing both states for prioritizing industrial interests over ecological sustainability. This article calls for the revitalization of Mahanadi, urging political and environmental reforms to prevent further degradation of the river's natural flow and ecological balance. Choudhury et al. (2023) provided a detailed analysis of water demand, land use changes, and pollution in the Mahanadi River Basin. This highlights the impact of human interventions, such as dams and reservoirs, on basin hydrology and stresses the need for improved water management practices in Chhattisgarh and Odisha. The Mahanadi River water conflict is deeply rooted in political power struggles, where both states have been accused of using river water for industrial benefits at the cost of agricultural and ecological needs (Sahu 2024).

The Government of Odisha announced its commitment to exploring all options to resolve the ongoing Mahanadi water dispute with Chhattisgarh (New Indian Express 2024).

12.2.1 Description

A Social Rift Rivers are the epicenters of human civilization. The historic Mahanadi binds neighboring people socially and culturally. Revered by local populations, rivers are not only vital sources of life but also hold deep religious significance. It plays a pivotal role in religious ceremonies, with its waters deemed sacred and used for rituals, especially in the worship of Lord Shiva, as numerous temples dedicated to the deity line its banks. Additionally, the river is integral to Jagannath culture, as historical accounts suggest that the deity Nilamadhaba, worshipped by Biswabasu, was transported along this river. Historians believe that the Sabar Narayan shrine marks the site where Biswabasu once worshipped Lord Nilamadhaba, further cementing the river's cultural and religious importance to both states.

The communities in Odisha and Chhattisgarh have a shared emotional and cultural connection with Mahanadi, reflected in folklore and songs celebrating its grandeur. The people of both states, particularly tribal groups, share common customs and often reside on either side of the state borders, seamlessly integrating their livelihoods across regions. This deep connection, however, has been strained by disagreements between the two state governments over the construction of dams in Chhattisgarh, which has led to protests and disruptions in Odisha.

The Mahanadi River Basin spans a vast area of Chhattisgarh and Odisha, with a catchment covering approximately 120,500 square kilometers. The river stretches 851 kilometers, with 357 kilometers passing through Chhattisgarh and 494 kilometers through Odisha, eventually draining into the Bay of Bengal. The basin is geographically significant and bordered by the Central Indian hills to the north, the Eastern Ghats to the south and east, and the Maikala range to the west. The Hirakud Dam, a prominent feature of the river, highlights its importance for water management. The Mahanadi River ranks as the second most significant peninsular river, following the Godavari River, in terms of water potential and flood capacity.

The state-wise distribution of the drainage of the Mahanadi River Basin

State	The drainage area (sq. km.)
Bihar	635
M.P. (now Chhattisgarh)	75,136
Odisha	65,580
Maharashtra	238
Total	141,589

Source: Government of India, Indian Irrigation Commission (1972): Report of the Irrigation Commission, Vol. I, Ministry of Irrigation and Power, New Delhi

12.3 Historical Background

On 27 April 1983, a meeting was held in Bhubaneswar among the then Madhya Pradesh Chief Minister Arjun Singh and his Odisha counterpart Janaki Ballav Patnaik, along with other ministers and officials from both states. Chhattisgarh was then part of Madhya Pradesh. Both states reached agreements on several river dams and power projects, i.e., Upper Jonk, Lower Jonk, Ib, Sapanai, Ongo, Kurnala, Lower Kolab, Jira, and Sahajbahal. Odisha and Madhya Pradesh agreed to form a "joint control board" to look into these interstate irrigation and power projects.

The Hirakud Dam of Odisha, which is constructed on the Mahanadi River, is one of the largest dams in the world. It is truly an engineering marvel designed by engineer M. Visvesvaraya. However, after the Hirakud Dam project, Madhya Pradesh constructed several irrigation projects, which directly obstruct water flow to the Hirakud Reservoir. Despite an agreement for a joint control board for equitable Mahanadi water distribution, this proposal has remained inactive for years (Government of Odisha 2004).

With time, a new state, Chhattisgarh, was formed on 1 November 2000. After 16 years of its formation, the dispute over the Mahanadi River began in 2016 when Odisha raised concerns over Chhattisgarh's construction of barrages and dams. These structures, intended for irrigation and industrial use, significantly reduced the water flow downstream, affecting agriculture and the drinking water supply of Odisha. The absence of an interstate agreement further complicated the situation, leading to the formation of the "Mahanadi Water Dispute Tribunal" in 2018.

The Mahanadi River dispute has its origins in the geographical and socioeconomic characteristics of the river basin. The river begins in Sihawa Mountain in Chhattisgarh, where it flows through Odisha before reaching the Bay of Bengal. The basin sustains a variety of ecosystems and supplies water for drinking, agriculture, and industry. However, the increasing demand for Mahanadi's resources has sparked conflicts over water distribution.

12.4 Industrial Impacts

Both Odisha and Chhattisgarh diverted substantial amounts of water from Mahanadi for industrial purposes. This diversion has led to severe water shortages, particularly in the lower basin of Odisha. The construction of six industrial barrages in Chhattisgarh exacerbates this issue, restricting water flow and impacting agriculture and local livelihoods in Odisha. The industrialization of the Mahanadi River Basin has been a significant factor in water disputes. Chhattisgarh, with its rich mineral resources, has experienced rapid industrial growth, leading to increased water demands. The construction of barrages and dams in Chhattisgarh has been driven by the need to supply water to industries, including power plants and steel factories. This has resulted in reduced water availability for downstream users in Odisha, where agriculture is a major economic activity.

The construction of dams and barrages by the Chhattisgarh government has significantly reduced the water flow to Odisha, the lower riparian state. In a legislative address, Odisha's former water resource minister, Raghunandan Das, noted that Chhattisgarh built 516 barrages in the upper Mahanadi catchment area, cutting water inflow to the river by 37% during the non-monsoon period from 2005 to 2016. Since 2016, Odisha and Chhattisgarh have been locked in a prolonged dispute over the construction of six industrial barrages—Kalma, Saradi, Basatpur, Mirauni, Sheorinarayan, and Samoda. These barrages are central to the conflict between the two states and have further reduced the downstream water flow (Mohanty 2021).

12.5 Political Manipulation of the Dispute

The Mahanadi water dispute has also become a political tool, with ruling parties in both states leveraging the issue for electoral gains. The ruling party of Odisha, Biju Janata Dal (BJD), and the Bharatiya Janata Party (BJP) of Chhattisgarh both used the conflict to gain support, further complicating the resolution process. However, the BJP is now ruling in both states. Therefore, the challenge is very difficult for both state governments to address the problem without hampering the sentiment and self of the citizens of both states.

The politicization of the Mahanadi River dispute has added another layer of complexity to the conflict. Political parties in both states have used the issue to support and gain electoral advantage. The BJD in Odisha accused the BJP-led government in Chhattisgarh of diverting water to benefit industries at the expense of farmers and communities in Odisha. Moreover, the BJP has defended its actions, citing the need for industrial development and economic growth. This political posturing has made it difficult to achieve a cooperative and mutually beneficial resolution to the dispute. The previous BJD government of Odisha accused Chhattisgarh of constructing dams and barrages without consultation, violating the principles of equitable water sharing. The Odisha government has also alleged that these constructions have led to reduced water flow downstream, adversely affecting agriculture and the drinking water supply in the state. However, the Chhattisgarh government has argued that construction is necessary for irrigation and industry purposes, blaming Odisha by saying that Odisha has not fully utilized its own water resources efficiently.

In 2023, the BJP again came to power in Chhattisgarh. This back-and-forth has created a political deadlock, with the rigidness of both states in their opinion. The lack of a formal interstate agreement on water sharing has exacerbated the situation, making it difficult to reach a conclusion.

The role of the central government in Mahanadi disputes has also been influenced by political dynamics. The establishment of the Mahanadi Water Disputes Tribunal in 2018 was a response to the demand of Odisha for federal intervention. However, the tribunal's progress has been slow, and its tenure has been extended multiple times, reflecting the complexities involved in resolving the dispute.

With the BJP now ruling in both states, the political challenge is more nuanced. The central government has a vested interest in ensuring that the dispute does not escalate, as it could impact the political fortunes of the BJP in both states. However, achieving a resolution that satisfies both states without compromising the interests of either remains a daunting task. Recent attempts at resolution have included joint meetings between officials from both states and attempts to create a framework for cooperation.

12.6 Ecological and Community Effects

The continuous construction of dams and barrages has led to significant ecological damage. Reduced water flow has degraded the natural ecosystem of Mahanadi, affecting flora and fauna. Moreover, industrial pollution further contaminates rivers, exacerbating environmental concerns. Odisha-based engineer Lalit Mohan Patnaik noted that Chhattisgarh has a 10% greater catchment area in the Mahanadi River Basin than Odisha does. While Chhattisgarh harvests approximately 30,000 million cubic meters (MCMs) of water, Odisha harvests 27,000 MCMs. Odisha is concerned that once all 2000 dams and barrages in Chhattisgarh become fully operational, Mahanadi could be transformed into a long reservoir with the capacity to store 829 MCM of water during the non-monsoon season (Prava 2021).

According to Patnaik, an independent environmental service professional, the water flow into the Hirakud Dam has already been reduced to one-third during the non-monsoon period, and once all of Chhattisgarh's dams and barrages become operational, the inflow will nearly cease. A 2018 study, entitled *Mahanadi: Coal Rich, Water Stressed*, conducted by renowned water expert Ranjan Panda, known as the "Mahanadi River Waterkeeper," also highlights the severity of the issue. The data submitted by the Odisha government reveal a grim scenario, showing that while Chhattisgarh, which holds 90% of the catchment area, currently sends 19,237 million cubic meters (MCMs) of water to Hirakud, Odisha's future demand will rise to 23,651 MCM. This resulted in a shortfall of 4414 MCM from Chhattisgarh's catchment area. Moreover, Hirakud received 2119 MCM from Odisha, which holds only 10% of the catchment. Once Chhattisgarh fully utilizes its projects, the total inflow to Hirakud will drop to 2119 MCM, far below the required 18,175 MCM, making the dam virtually nonfunctional (Prava 2021).

The ecological impact of the Mahanadi River dispute is profound. The reduction in water flow has led to the degradation of wetlands, loss of biodiversity, and disruption of the natural hydrological cycle of the river. The construction of dams and barrages has altered the flow regime of Mahanadi, affecting fish migration and spawning. Additionally, industrial pollution has contaminated rivers, posing a threat to aquatic life and human health. The cumulative impact of these changes has led to a decline in the ecological integrity of the river.

12.7 Effect on Local Populations

Reduced water flow has dire consequences for riparian communities, particularly farmers and fishermen. In Odisha, many farmers have faced crop failure due to inadequate irrigation, while fishermen have experienced a decline in fish populations, which threatens their livelihoods. Additionally, water scarcity has led to severe drinking water shortages, affecting public health.

The social impact of the Mahanadi River dispute is equally significant. Farmers in Odisha have struggled with water scarcity, leading to reduced agricultural productivity and economic hardship. The decline in fish populations has affected the livelihoods of fishermen, who rely on rivers for their income. Water scarcity has also led to conflicts over water allocation, with communities competing for limited resources. The health of riparian communities has been compromised, with drinking water shortages affecting public health.

Between 2005 and 2007, the irregular water supply to the Sason canal, which is situated on the left bank of the Hirakud Dam, disrupted irrigation and affected the livelihoods of approximately 60,000 farming families in the canal's command area. Rather than prioritizing water for agriculture, the Odisha government focused on industrial development. Several industries, including Vedanta Alumina, Aditya Alumina, and Hindalco, constructed pipelines and intake wells within the reservoir to extract water. This industrial preference significantly decreased the amount of water available for irrigation, prompting approximately 30,000 farmers to protest at the Hirakud reservoir in Sambalpur in 2007 against the increasing diversion of Mahanadi water for industrial use. The government responded by promising to prioritize water for farmers and ensuring that industries would receive only surplus water. However, farmers' struggle for water access continues (Purohit 2016).

"Mahanadi flows 72 km by the Athamalik subdivision (the longest for any subdivision in the state), which is perennially prone to drought. It has been more than five decades since the government promised an irrigation project. Providing water to farmers has never been a priority for the Odisha government. Most political parties in the country say that the water belongs to farmers, but none take a stand against industries. Why point gun at farmers when the culprit is someone else?" asks the senior farmer leader Ashok Pradhan (Purohit 2016).

12.8 Attempts at Resolution: Mahanadi Water Dispute Tribunal

The Mahanadi Water Dispute Tribunal, established in 2018, aims to resolve conflicts by inspecting river basins and assessing water availability and usage. However, progress has been slow, with the tenure of the tribunal extended to 2025. Despite these efforts, the lack of cooperation between the states continues to hinder effective resolution.

The Mahanadi Water Dispute Tribunal was established to address the complex issues surrounding river disputes. The mandate of the tribunal includes examining river basins, assessing water availability, and making recommendations for water allocation. However, progress has been slow, with delays in data collection and analysis. The lack of cooperation between Odisha and Chhattisgarh has hindered the efforts of the tribunal to reach a resolution. The extension of the tenure of the tribunal until 2025 reflects the ongoing challenges in resolving the dispute.

12.9 Community Organization and Advocacy Campaigns

Various activist groups, such as Mahanadi Bachao Andolan, have emerged as advocates for the protection and equitable water distribution of Mahanadi. These movements emphasize the need to prioritize ecological and community rights over industrial demands, pushing for sustainable management of the river basin.

Civil society and activist movements have played crucial roles in advocating for the protection of the Mahanadi River and equitable water distribution. Groups such as Mahanadi Bachao Andolan have raised awareness of the environmental and social impacts of water disputes. These movements have called for greater transparency in water management and the prioritization of ecological and community rights over industrial demands. By mobilizing public support and engaging with policymakers, civil society groups have pushed for sustainable management of the river basin.

12.10 Suggested Framework for Settlement: Interstate Collaboration

A cooperative framework involving both states is essential for resolving the Mahanadi dispute. This framework should include transparent data sharing, joint water management strategies, and equitable distribution agreements to ensure that the needs of both states are met without compromising the ecological health of Mahanadi.

Interstate cooperation is essential for resolving the Mahanadi River dispute. A cooperative framework should include transparent data sharing, joint water management strategies, and equitable distribution agreements. By working together, Odisha and Chhattisgarh can ensure that the needs of both states are met without compromising the ecological health of the river. This approach requires a commitment to dialogue and collaboration, with both states recognizing the importance of sustainable water management.

12.11 Regulatory and Policy Changes

National reforms are crucial to securing community rights over river resources and acknowledging the ecological rights of the Mahanadi River. Strengthening the "Interstate River Water Disputes Act" and enforcing water-sharing agreements rigorously can help reduce conflicts and encourage sustainable water management. Legal and policy changes are essential for addressing the core issues of the Mahanadi River dispute, with national-level reforms focusing on community rights and the river's ecological well-being.

12.12 Ecological Revitalization

Revitalizing the ecology of the Mahanadi River is crucial for long-term sustainability. This involves regulating industrial pollution, restoring natural water flow, and implementing conservation projects to protect the biodiversity of the river. Collaborative efforts between the government, civil society, and local communities are vital for successful ecological restoration.

Ecological revitalization is essential for the long-term sustainability of the Mahanadi River. This involves regulating industrial pollution, restoring natural water flow, and implementing conservation projects. By addressing the root causes of ecological degradation, it is possible to restore the health of Mahanadi and protect its biodiversity. Collaborative efforts between the government, civil society, and local communities are vital for successful ecological restoration. By working together, stakeholders can ensure the long-term health and sustainability of the Mahanadi River.

12.13 Conclusion

The Mahanadi River dispute reflects the complex interplay of politics, industry, and environmental concerns in interstate water conflicts. Addressing this issue requires a multifaceted approach, and the states of Odisha and Chhattisgarh can work toward resolving the Mahanadi conflict and ensuring its health for future generations. As per the former Prime Minister Shri Atal Bihari Vajpayee, "Governments will come and governments will go, but the country will always remain." Therefore, different political parties forming the governments in both states should not affect the equitable distribution of water. The stakeholders from both states should not suffer because of this issue. Both state governments must act while keeping this in mind.

References

Choudhury SR, Singh AK, Mahato A, Anand A, Chandrakar AK (2023) A comprehensive review of the Mahanadi river basin in Chhattisgarh, India. Ecol Environ Conserv 29(4):147–157. https://doi.org/10.53550/EEC.2023.v29i04.031

Government of Odisha (2004) Orissa: state water plan 2004, interstate agreements. Department of Water Resources

Heinrich-Böll-Stiftung (2018) Mahanadi: Coal rich, water stressed. https://in.boell.org/sites/default/files/uploads/2019/04/report_mahanadi_water_dispute_2018_final.pdf

Mahapatra SK (2005) Water governance: a historical understanding of Mahanadi River Basin, Orissa (MIDS Working Paper No. 193). Madras Institute of Development Studies

Mohanty D (2021, December 10) Barrages on upper reaches of Mahanadi by Chhattisgarh has choked inflow to Odisha, alleges Minister. Hindustan Times

New Bharat Express (2024, June 27) Odisha government will explore all possibilities to resolve Mahanadi row: Pujari. The New Bharat Express. https://www.newBharatnexpress.com/states/odisha/2024/Jun/27/odisha-government-will-explore-all-possibilities-to-resolve-mahanadi-row-pujari

Prava P (2021, August 9) Can Odisha, Chhattisgarh move beyond conventional approaches to save Mahanadi? Hindustan Times

Purohit M (2016, November 3) Maha battle over water. Bharat Water Portal. http://www.Bharatwaterportal.org/articles/maha-battle-over-water

Ratha KC (2019) Revisiting the Mahanadi water dispute discourse: a reflection of diverse perspectives. Maharshi Dayanand University Research Journal ARTS 18(1):103–113

Sahu M (2024) The Mahanadi River water dispute discourse: an amalgamation of politics, power, and privatization. J Stud Soc Sci Humanit 8(2):232–245

Swain M (2017) Resolving the Mahanadi water conflict. Econ Polit Wkly 52(51):23–26

Water Conflict Forum (2017) Integrated water management of the Mahanadi basin. Water conflict forum. https://waterconflictforum.org/lib_docs/Integrated-Management-of-the-Mahanadi-basin.pdf

The manufacturer's authorised representative in the EU is Springer Nature Customer Service Centre GmbH, Europaplatz 3, 69115 Heidelberg, Germany. If you have any concerns regarding our products, please contact ProductSafety@springernature.com

Printed and bound by CPI Group (UK) Ltd, Croydon, CR0 4YY
01/10/2026
02225945-0017